1. Introduction

Global Horizons is the AF vision for global science and technology (S&T) spanning air, space, cyberspace; command and control (C2); intelligence, surveillance and reconnaissance (ISR); and mission support to address current and future threats. *Global Horizons* focuses on game changing S&T capabilities in the near, mid and far term that will advance the survivability, affordability, and effectiveness of AF global operations by leveraging global industrial sectors.

> *"The globalization of information technologies continues to fuel advanced military research and development abroad. Consequently, in some areas the U.S. is working harder to sustain more narrow military advantages."*
>
> **Honorable Michael Donley**
> **Secretary of the Air Force**

1.1 Motivation

Air Force systems are increasingly dependent upon the global domains for both mission enablement and mission delivery. Simultaneously, global domains (air, space, cyberspace) are an increasingly congested, competitive and contested environment. While domestic fiscal constraints drive a quest for

> *"The product that we provide for the nation is Global Vigilance, Global Reach, and Global Power"*
>
> **Gen Mark Welsh**
> **Chief of Staff, U.S. Air Force**
> **21 February 2013**

efficiency and economy, the global industrial sectors provide a unique opportunity to leverage $1.49 trillion in global R&D investment (Battelle 2013). We are challenged by a limited future supply of domestic science, technology, engineering and mathematics (STEM) graduates and also by the speed of attacks and velocity of threat evolution. Accordingly, it is imperative that we focus our investments on those areas most likely to return game changing capabilities that sustain our global advantage. Game changers provide ten to one hundred times improvements in efficiency or effectiveness.

1.2 Vision

The Air Force *Global Horizons* S&T vision aims to achieve "sustained global advantage that ensures *Global Vigilance, Global Reach and Global Power* in, through and from air, space and cyberspace." Each of these words bear important meaning. "Sustained" means ensuring

> **Global Horizons Vision**
> **Sustained global advantage that ensures Global Vigilance, Global Reach and Global Power in, through and from air, space and cyberspace.**

operations in spite of vulnerabilities in militarily, economically, and politically contested environments. The AF interest in "global" operations spans research, development, acquisition, and employment. The "advantage" the AF seeks is a readiness, robustness, resiliency and responsiveness edge over our adversaries to ensure operational superiority in spite of complex, constrained, competitive, congested and contested environments. Finally, the AF requires superiority "in, through and from" full spectrum operations across air, space and cyberspace to support the joint and coalition fight. This vision is aligned with the AF heritage of focusing on strategic, global engagement. This means a focus on *Global Vigilance* through global persistence and awareness, on *Global Reach* via global access, speed, and stealth, and on *Global Power* via global, integrated, cross domain effects. The latter implies careful synchronization of AF, joint and international partner actions across air, space, land, sea and cyberspace.

1.3 Alignment

As illustrated in Figure 1.1, *Global Horizons* leverages and flows naturally from the White House *National Security Strategy,* Department of Defense (DoD) strategy, Air Force strategy and doctrine, strategic studies by the Air Force Scientific Advisory Board as well as the *Air Force Science and Technology Plan*, *Technology Horizons* (2010), *Energy Horizons* (2011), and *Cyber Vision 2025* (2012). The formulation of *Global Horizons* carefully considered AF doctrine and vision for *Global Vigilance, Global Reach, and Global Power*, joint, interagency, combatant command (CCMD) and major command (MAJCOM) requirements, the AF *Global Partnership Strategy*, and the 12 AF core function master plans (CFMPs). Maintaining a focus on AF operations, the study explored opportunities in global industrial sectors including transportation and logistics, manufacturing and materials, communications, information technology and financial services, energy, health care and pharmaceuticals, and education and training.

Figure 1.1: Strategic Alignment of *Global Horizons*

1.4 Methodology

The *Global Horizons* study was guided by a three-star governance team and an enterprise-wide set of key AF stakeholders (See Section 18). It was organized into core functions and global industrial sector panels in each of the areas shown on the right side of Figure 1.1, collaboratively partnering senior experts and leaders from MAJCOMs, Air Force Research Laboratory (AFRL), product centers, operational units, and Headquarters Air Force. Additionally, national, DoD, and AF strategy and policy provided guidance for areas of focus. To engage external expertise, a public request for information (RFI) and area- and sector-focused summits resulted in the consideration of hundreds of detailed studies, concepts and technologies from across the nation and abroad as exemplified in Figure 1.2. The core function and global industrial sector distribution of written responses to the RFI is shown in Figure 1.3.

Photo Removed Due to Copyright Restrictions

Figure 1.2: Extensive Subject Matter Expert Engagement

Team Members made focused site visits across the United States, United Kingdom, Germany and Australia. Multiple subject matter expert workshops/summits were held at major AF installations and included expert participants from industry, academia, government, national laboratories, and federally funded research and development centers (FFRDCs). Expert teams (see Section 18) incorporating operational, technical, and industrial sector experts assessed the very best of identified ideas and technologies, forecasted capabilities, and created an S&T roadmap for the near, mid and far term for each area. A senior independent expert review group (see Section 18) reviewed the results in two major reviews at the Pentagon which were assessed by the senior governance council and approved by AF leadership. Given the dynamicity, complexity, and strategic role of global S&T, AF engagement will require continued monitoring, planning and refinement.

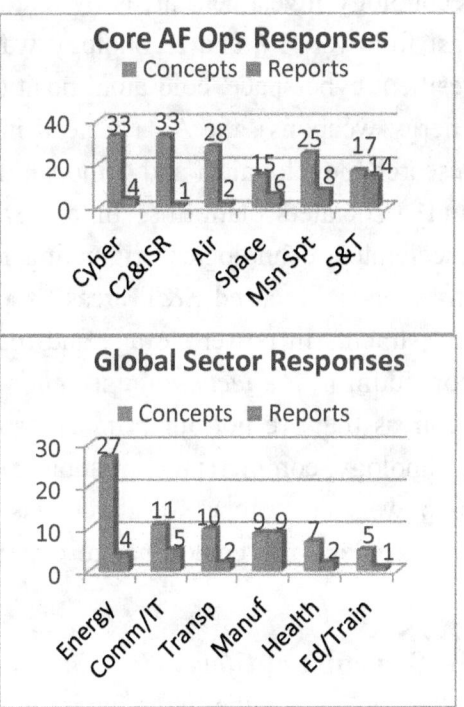

Figure 1.3: RFI Responses

1.5 S&T Partnerships

Given limited AF resources and significant global investment by others, the AF S&T approach is to maximally leverage the knowledge, capabilities, and investments of our sister services, departments, national laboratories, industry and industrial consortia, utilities, FFRDCs, universities, and international partners as illustrated in Figure 1.4. This approach allows the AF to preserve resources and focus investments on AF unique systems and core functions.

Photo Removed Due to Copyright Restrictions

Figure 1.4: Partnerships

1.6 S&T Roles: Lead, Follow, Watch

To clarify partnerships, roles, and responsibilities, *Global Horizons* articulates priority technology investment areas by distinguishing among three key roles: technology leader (L), fast follower (F), and technology watcher (W). In a *technology leader* role (e.g., trusted and resilient cyberspace, cold atom position, navigation and timing (PNT), hypersonic and directed energy weapons), the AF is a lead investor and creates or invents novel technologies through research, development and demonstration in areas that are critical enablers of AF core functions and associated platforms. In a *fast follower* role, the AF rapidly adopts, adapts, and/or accelerates technologies originating from external organizations who are leaders and primary investors in focused S&T areas as part of their core functions (e.g., Department of Energy investments in power storage and management, commercial investments in high performance computing). In a *technology watcher* role, the AF uses and leverages others' S&T investments in areas that are not our primary or core functions (e.g., commercial commodity information technology, commercial communications, manufacturing technology, critical infrastructure such as power and water). Roles were assigned using the consensus of small groups of experts and stakeholders and could change depending upon resource, operational priority, or technology changes.

1.7 Structure of *Global Horizons* Document

After forecasting the future environment and threat, *Global Horizons* addresses strategic trends, threats and game changing opportunities in the core AF operational areas of air, space, cyberspace, C2, ISR, and mission support. The document then similarly addresses strategic trends, threats and game changing opportunities in the global industrial sectors of manufacturing and materials, transportation and logistics, energy, communications, information technology and financial services, health care and pharmaceuticals, and education and training. The document concludes by recommending a way forward. A separate Appendix provides detailed discussion and justification of trends, game changers, findings and recommendations. This includes detailed technology roadmaps prioritizing where the AF should lead, follow, and watch in the near, mid, and far term.

2. Future Environment

Figure 2.1 characterizes key global forces, global industrial sectors, and the global domains of air, space, and cyberspace to set the stage for forecasting futures. All of the domains in which the AF operates are contested, congested, and competitive. Global communications, conveyance, and commerce depend upon freedom of action in these global domains which are increasingly congested, competitive, and contested. Moreover, the globalization of the industrial base, which will be addressed in further detail in subsequent sections, is both a strategic threat and opportunity for the AF. For example while the AF is threatened by increasing foreign dependency and possibility for surprise, there are increasing opportunities to diversify supply, leverage investments, and partner to accelerate progress.

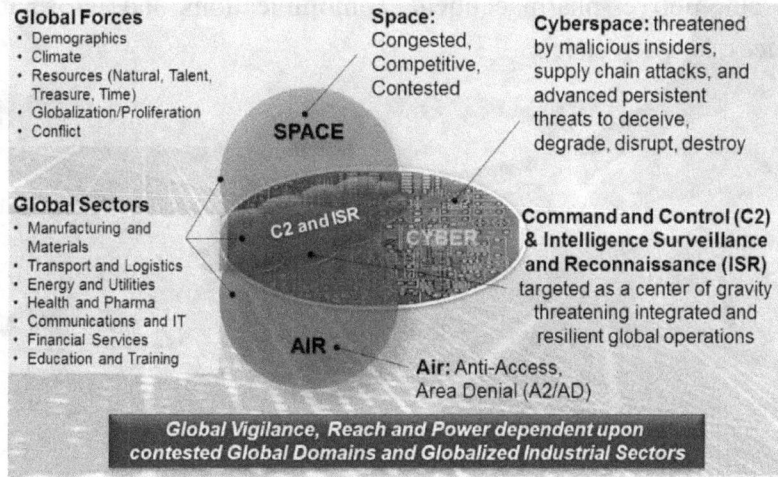

Figure 2.1: Global Domains and Global Industrial Sectors are Increasingly Strategic

2.1 Strategic Trends

Figure 2.2 illustrates key demographic, economic, resource, technological, threat and investment trends that are shaping the future environment (See Appendix for sources). By 2025, we forecast that 56% of the world's eight billion people will reside in Asia—making it an attractive commercial market for advanced information technologies. As is reflected in the comparative growth and national focus, by 2025 China will produce more than double the number of computer science doctorates as the US. By 2050, the world's population will grow to over nine billion and be increasingly urban (growing from 50% to 70%), middle class (from 50 to 65%), and older (from 31 to 41 years on average, but unevenly distributed with those over 60 years of age doubling from 10% in 2000 to 21.5% in 2050). Bulging population will place increased strain on limited resources For example, at current production and consumption rates, the world supply of Indium (used in WWII to coat bearings in high-performance aircraft and now in liquid crystal displays and touchscreens) is expected to last only eight years. Limitations of some critical resources (e.g., water, energy, minerals) could drive future conflict. Combined temperature and humidity increases are expected to drive more frequent severe climate events. Explosive growth in communications and computing will accelerate progress in all sectors; however, exponential increases in malware will threaten increasingly dependent infrastructure, systems and services. A doubling of foreign satellites on orbit by 2033 will provide new challenges in space. However, there are positive aspects of this challenging future. For example, transportation costs, desire for local, rapid market access, and new technologies such as additive manufacturing will reverse some offshoring of manufacturing. Accelerating technology advances and adoption will create new wealth and the growing global middle class will demand higher quality education, housing, health care, environment, and governance, all of which will drive security, stability and prosperity. Moreover, as the public and private sector increase the current $1.4 trillion investment in wealth- and security-producing research and development, there will be numerous opportunities to leverage multi-trillion dollar annual markets in industries such as automotive, pharmaceutical, communications and information technology (IT), financial services, and aerospace.

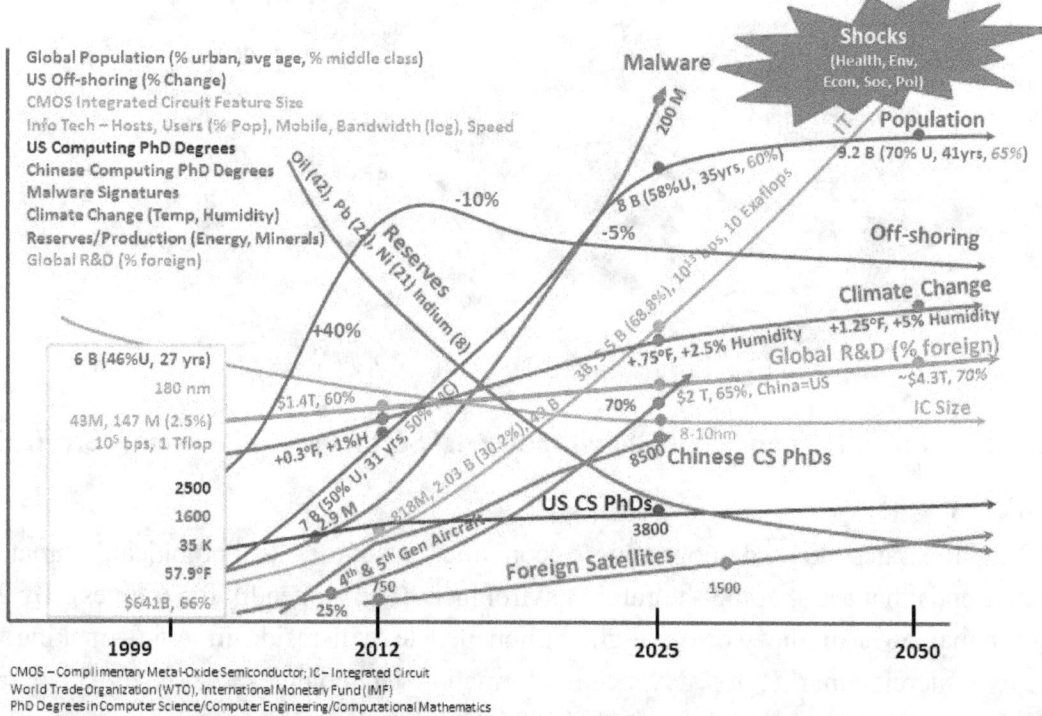

Figure 2.2: Strategic Trends through 2050

2.2 Global Threats

Knowledge of and access to technology is increasing each year. While access to information is important, the ability to understand and apply it to problems, whether military or civil, is increasingly critical. The U.S. has enjoyed an advantage in the development and application of new technology for decades but that advantage is shrinking. The greater proportion of global R&D investment and ever-increasing percentage of STEM graduates outside the U.S. support this trend. Insufficient domestic STEM graduates threaten our ability to conduct national security R&D and manage S&T programs. There are several technological areas of relevance to the AF where international investment creates the potential for partnership (e.g., logistics R&D in Australia, graphene research in the UK, manufacturing technology in Germany, biofuels in Brazil, robotics in Japan/Korea). Additionally, global macro-trends such as population growth, climate change, and competition for shrinking natural resources will have effects on technology but are beyond the scope of this study.

There are numerous global trends that do present potential threats to the AF through 2025. This section will not list them all but will address some of the top threats in each area as well as three technologies which, in the hands of our adversaries, continue to be of particular concern because of their potential impact across multiple AF core functions: electronic warfare, directed energy weapons, and cyber.

- *Air* – Remotely piloted aircraft (RPA) development will increase significantly in the coming decade. While most systems deployed or in development today are ISR-related, significant investments are being made to develop combat RPA with some potentially capable of delivering weapons of mass destruction (WMD) (chemical/biological/ nuclear). Detection and defeat mechanisms will need to be developed. The deployment and proliferation of 5th generation fighters will also be a concern. By 2025, 70% of foreign combat air forces will be comprised of modern 4th or 5th generation aircraft.
- *Space* – Counters to our space advantage are being developed and proliferated worldwide. Jamming of communications and PNT will become widespread with the ability to disrupt ISR operations and achieve physical destruction of space assets becoming more prevalent over the next decade. Attribution and locating threats will remain a challenge.
- *Cyberspace* – Perhaps no technological area has greater potential to cause an asymmetric advantage in the future battlespace than information technology and cyberspace. Malware threats are increasing in complexity and number and can be embedded and lay dormant in existing systems until activated or can be targeted against a specific system or capability. The dependence of AF systems on cyber, the relatively low cost and speed of "weapons" development, and the difficulty of attribution will make cyber attacks an increasingly attractive option for all U.S. adversaries.
- *Anti-Access, Area-Denial* (A2/AD) – Certain peers, near peers, and other adversaries will employ anti access strategies which will effect traditional U.S. basing options. Additionally, China and Russia will improve their cross-domain capabilities. China, in particular, will place considerable effort in integrating and synchronizing air, space, ground, maritime, missile and cyber capabilities by 2025. Ballistic/cruise missiles,

combined with cyber operations, could complicate U.S. forces' ability to enter and operate in theater.

- *Electronic Warfare (EW)* – Foreign advances in EW will increase our challenges as digital systems allow adversaries to rapidly reprogram and modernize their weapons systems. The AFs archaic process of collecting, analyzing, developing counters, and reprogramming is inadequate for today's digital combat environment.

- *Directed Energy Weapons (DEW)* – Advances in R&D will accelerate the development and deployment of DEWs (high powered lasers and microwaves) by our adversaries which can disrupt or deny air, space, and cyber operations.

- *WMD* - WMD are a major consideration the AF must address, but as part of the national response which is beyond the scope of this report. S&T considerations were recently addressed in the 2013 *Defense Science Board Study on Technology and Innovation Enablers for Superiority in 2030.*

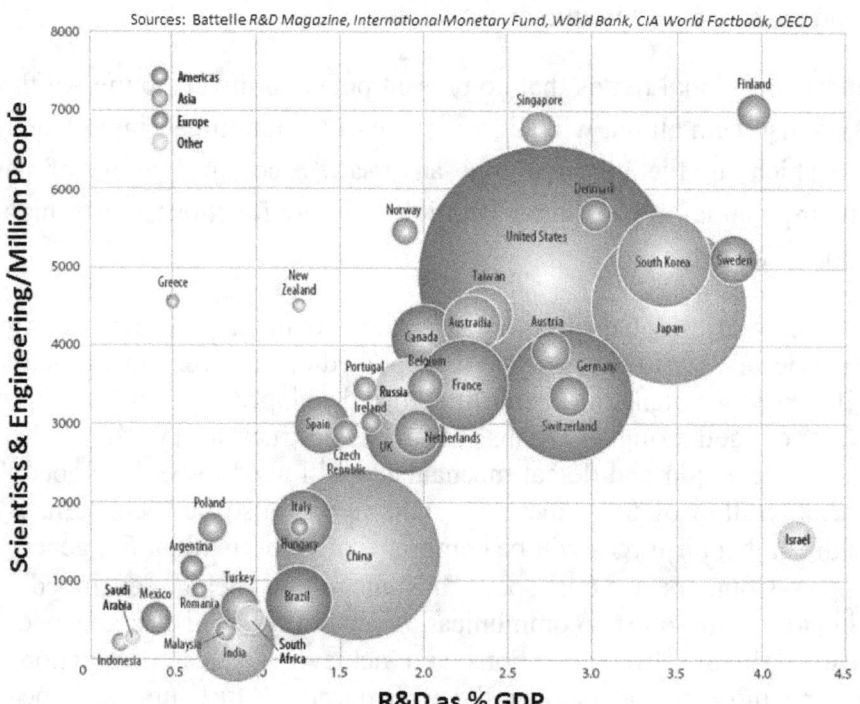

Figure 2.3: R&D Expenditures Global Industrial (2010)

2.3 Global Opportunities

Figures 2.3 and 2.4 decompose the $1.4 trillion of annual investment in R&D distributed across countries, continents, global industrial sectors and global companies. Figure 2.3 displays global investment distributed among the Americas, Europe, and Asia, with Asia growing proportionally over time. Figure 2.4 illustrates the scale of the global pharmaceutical, telecommunications, information technology, and automotive industries, underscoring the opportunity for investment partnership and leverage. Recognizing associated economic, social, and security benefits, leading industrialized nations expend between 2.5% to 3.5% of GDP on research and leading industries expend between 5% and 20+% of revenues on research.

Subsequent sections of the report identify specific opportunities where the AF can engage in partnerships and leverage these external investments.

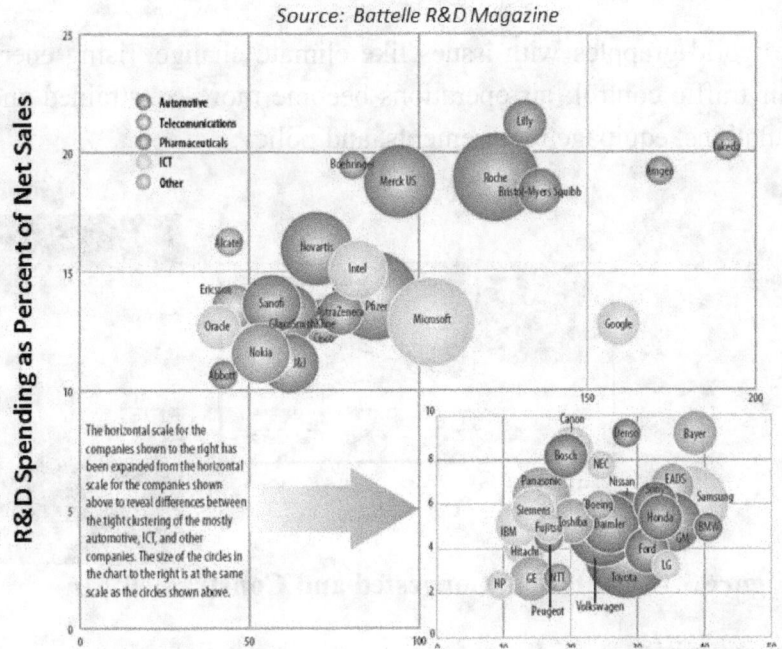

Figure 2.4: R&D Expenditures by Global Industries (2010)

Given these global threats and trends, the remainder of the document provides more detailed findings and recommendations for game changing opportunities within AF core functions as well as global industrial sectors.

3. Air Domain

3.1 Trends

Air superiority is challenged by several strategic trends as illustrated in Figure 3.1, including:

Contested: Globalization, improving wealth of potential near-peers, and ready access to technology will shrink the time during which the U.S. enjoys technological advantages. Adversaries will rapidly gain advanced systems (5th generation fighters, new missiles, munitions, and DE capabilities) and learn how to employ them. We must plan and train to operate, survive, and execute missions in anti-access and area denial environments. If game-changers allow our forces definitive success in such environments, nuclear-armed adversaries may then compel the AF to rapidly shift to operations in a nuclear environment, yet another contested challenge.

Congested: While the AF manned airfleet will shrink slightly through 2027, RPA fleets and missions will grow significantly, with commensurate challenges in air safety, control, and cyberspace security. The US is being outpaced in military expenditure by the rest of world, which will likely lead to an increase in capability and diversity of international military air forces. Commercial aviation anticipates a doubling in air traffic volume, with a possible tripling in the Asia-Pacific region by 2030.

Constrained: As the world grapples with issues like climate change, rising energy costs, and paradigm shifts in air traffic control; air operations become more constrained and sculpted by costs, mandates, regulations, equipage requirements, and policy.

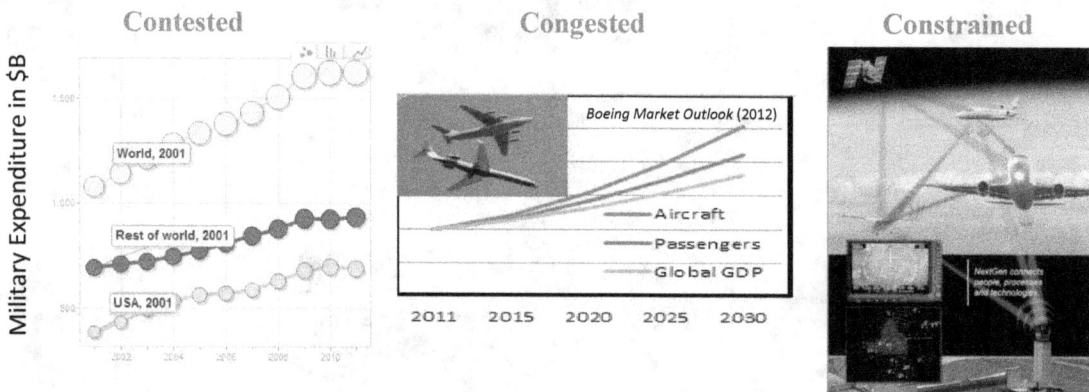

Figure 3.1: Contested, Congested and Constrained Air

3.2 Threats and Opportunities

The increasing average age of AF platforms allows adversary capability to approach ours and drives large sustainment costs of legacy fleets (+180% per decade). The gap between government and commercial air investment affects operating costs—military small lot buys of specialized platforms are expensive. Long acquisition times exacerbate these threats. Realistic threats need to considered at all stages of the acquisition process, beginning with requirements. Developmental risks emerge when test facility assets and ranges do not incorporate these threats and are left to deteriorate or close. AF work force demographics and supply/demand disconnects in the STEM workforce create inefficiencies.

Opportunities arise from technologies which satisfy mandate/regulation requirements, while simultaneously providing enhanced mission capability. For example, the C-5M has a new digital backbone, new avionics, and improved engines to address communications, navigation, and fuel efficiency requirements. As a result, on-wing time is up five-fold for the new CF-6 engines; individual sortie reliability rate is up over 10% resulting in number of maintenance delays per mission cut in half; and range has expanded 27%, allowing fewer enroute stops or aerial refuelings.

3.3 Game Changers

An overarching consideration in all potential technologies is system affordability. Although these game changing themes apply across the entire AF, their impact and development will likely first occur in RPAs and will reformat our current modes of operation. We should partner in development where there is overlap with commercial, joint, and coalition interests.

Autonomy/Fractionated² Systems/Distributed Decision Making: Achieving true autonomy has been elusive and challenging in air systems, however a confluence of improved computing power, cybersecurity, and connectivity offer transformational opportunities in autonomous systems (large and small platforms), distribution of functions across a set of "fractionated" systems (spatially separated functions, but joined in communication and operation), and the realization of distributed decision making based on mission requirements. In some cases, the autonomous systems will perform offensive operations, for these we must be aware of policy (DoDD 3000.09, *Autonomy in Weapon Systems*): "[a]utonomous and semi-autonomous weapon systems shall be designed to allow commanders and operators to exercise appropriate levels of human judgment over the use of force."

Speed: Prompt global strike and the ability to project power erodes as potential adversaries improve advanced anti-access strategies and area denial capabilities such as spectrum control and guided weapons. Swift, low observable, maneuverable, and agile systems are more survivable as they reduce exposure time and allow for quick response to known threats. Nothing moves faster than light, and advances in efficiencies, power levels, thermal management, and optics make directed energy weaponry a game-changing contender.

Advanced Aircraft Adaptive Architecture: Open architectures and modular components (sensors, seekers, etc.) will allow weapons systems to rapidly adapt to changing missions. Instantaneous connectivity and recognition of attached armaments provide a plug-and-play

---

² A fractionated system is a physically and functionally distributed system whose elements interact such as the internet. In contrast, a system like the F-35 or DSP is fully integrated. Fractionation can enhance survivability, graceful degradation, and reconstitution. Open standards and loose coupling can facilitate composition/decomposition and interoperability. Fractionation can be isomorphic (e.g., the GPS constellation) or polymorphic (e.g., SBIRS high and low).

approach. This allows easier system upgrades, mission-specific avionics and adaptable weapons system configurations, but could present new threat vectors unless cybersecurity is built in.

3.4 Recommendation

Conduct a series of flight tests, experiments, and challenges to demonstrate an effective, robust, partnership of manned and unmanned air platforms—validating key concepts of autonomy, fractionated systems, and distributed decision making in realistic threat and permissive environments. (OPR: AFMC; OCR: ACC, AMC, AFGSC, AFSOC) A stakeholder IPT should:

- Define and validate a methodology to measure key machine, human, and mission performance metrics.
- Select representative technologies (e.g., human-machine cognitive communications, plug-and-play avionics and armaments interfaces, trust in cyber systems) and mission functions for consideration.
- Generate an integrated roadmap for development, test, and exercises to verify savings and improvements in operational capability.

4. Space Domain

4.1 Trends

With the world-wide proliferation of space launch and small satellites, space is becoming increasingly congested, contested, and competitive as illustrated in Figure 4.1. A major trend is that virtually any country can procure launch services and easily access space. An exemplar of space congestion is the increasing amount of debris, consisting of tens of thousands of objects sized 10-cm or greater, with an estimated 100 million objects smaller than that. In the past ten years the number of objects tracked has grown by a third. Similarly, contested space is illustrated by the increasing vulnerability of our high-value space assets, growing cyber and physical threats. Finally, the overloaded electromagnetic (EM) communications spectrum, both in this country and internationally, reflects the competitiveness of space. The U.S. must move to increasingly higher frequencies (e.g. V-band, W-band, lasercom) in order to avoid the allocation difficulties present at all lower frequencies.

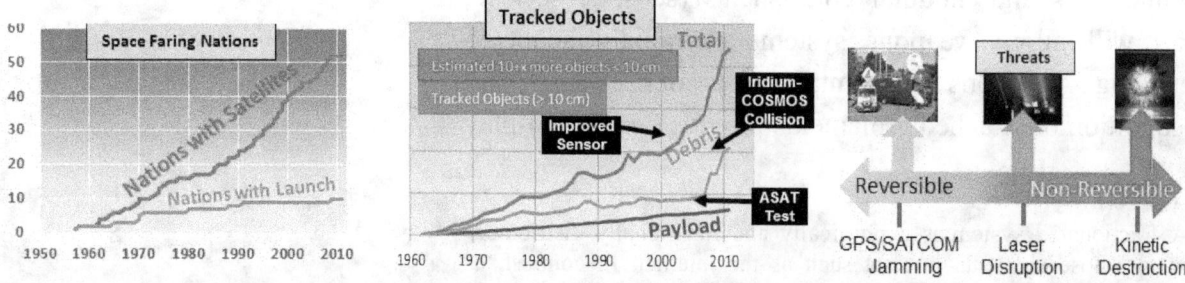

Figure 4.1: Competitive, Congested and Contested Space

4.2 Threats and Opportunities

There are clear threats to the U.S. space enterprise, including growth in space debris, space weather induced upsets, the increasingly easy access to space, and potential cyber/EW/kinetic attacks on our space and space-support ground assets. In addition, there are severe budgetary threats, including the current DoD acquisition and programming system that preserves large, legacy programs-of-record that increase costs of space assets, while discouraging the rapid insertion of capability that uses new advances in commercial technology. The EM spectrum allocation is severely constrained by S and L band openings for purchase by the commercial sector. All of this is exacerbated by the increasing capabilities of our adversaries because of STEM investments while at the same time there is declining interest in STEM by U.S. students.

However, these threats present opportunities for the AF to revamp the way we provide space services. For example, the contested space issues (cyber, EM spectrum) are opportunities for international cooperation to improve Global Positioning System (GPS) accuracy, develop protocols for cyber cooperation, and open up new EM spectrum for communications and control. We can revolutionize our space architectures by using hosted payloads and launching smaller, affordable, and fractionated satellites in disaggregated constellations, as well as implementing rapid, innovative acquisition practices that exploit the predicted rapid growth in space tourism and small launch vehicles. National competitions such as the X-Prize or the Defense Advanced Research Projects Agency (DARPA) Grand Challenge could energize students, and new technologies such as reconfigurable modules for spacecraft docking and servicing or carbon nanotube thrusters could generate considerable excitement.

4.3 Game Changers

Disaggregated Systems and/or Fractionated Satellites: Subject to affordability and architectural considerations, game changing approaches could make use of disaggregated systems and/or fractionated satellites to complement few, very large and highly capable legacy satellites to provide resilience, reduce vulnerability, and balance performance and cost effectiveness.

Small/Low-cost Launch: To efficiently exploit smaller platforms, the small, low-cost launch capability being developed by commercial industry provides a new paradigm for accessing space.

New Technologies: New technologies such as additive manufacturing in space (enabling on orbit construction and repair), combined with modular and open architectures can help realize low-cost satellites, and agile, reconfigurable space systems. Autonomous space systems and ground control would revolutionize space operations; but the biggest impact would come from increasing satellite power, persistence, and survivability to conduct ISR and other traditional air missions from the relative sanctuary of space.

4.4. Recommendations

- Pursue "Disaggregated" satellite constellations (OPRs: AFSPC, SMC)
- Utilize existing and emerging commercial launch operations for small payloads. (OPRs: AFSPC, SMC, AFRL)
- Redefine space acquisition in accordance with disaggregated satellites and inexpensive launch (OPRs: AFSPC, SMC; OCR: SAF/AQX) with a goal of greater than 10x cost reduction employing advanced technologies.
- Pursue AF "game-changing" technologies for space (OPR: AFRL) (e.g., adaptive manufacturing in space, lasercom and quantum computing, High Assurance Internet Protocol Encryptor (HAIPE) enabled satellites, autonomous operations (including ground), air-space integration).

5. Cyberspace Domain

5.1 Trends

Key trends contouring the future cyber environment include increased government use of commercial off the shelf (COTS), exponential growth of malware, increased use of cloud computing, and increased complexity of systems. New pieces of malware have increased more than tenfold from 9 million in 2007, to over 100 million in 2012, with over 200,000 new malicious programs registered each day. By 2015, experts predict that 20% of information will be processed and/or stored by the cloud. While information technology organizations continue to suffer the most significant increase in cyber exploitation, Figure 5.2 details the 882% increase in federal agency reported cyber incidents since 2006 (Source: United States Computer Emergency Readiness Team (US-CERT)). The complexity of the cyber world is evident by growing software sophistication and total worldwide data production from information systems (See Figures 5.1 and 5.3). An encouraging trend is the move toward ubiquitous encryption, a direct consequence of the desire to store sensitive data on untrusted assets, such as in the public cloud.

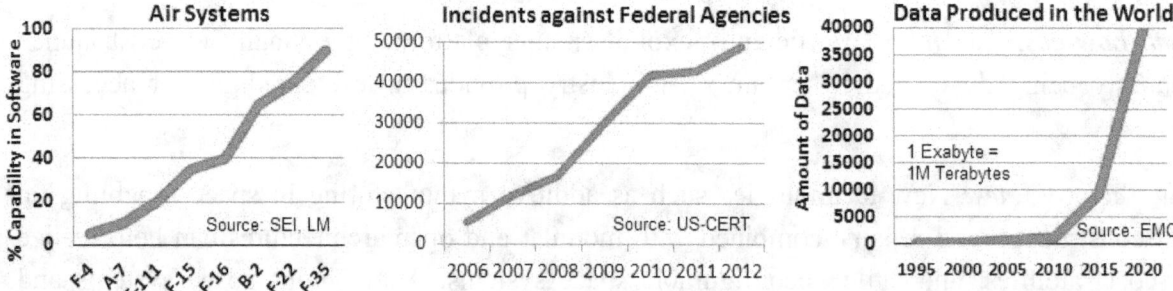

Figure 5.1: Percent of Aircraft System Capability in Software

Figure 5.2: Number of Reported Cyber Incidents against Federal Agencies

Figure 5.3: Total Worldwide Data Production

5.2 Threats and Opportunities

Critical threats to cyber operations include the vulnerability of systems and processes. The loss of IT supply chain integrity due to outsourcing to Asia and the presence of malicious insiders increases attack risk. In part due to growing software complexity, software vulnerabilities have increased. One estimate is that one vulnerability is introduced for every 1,000 lines of code (Perrin 2010). Supply chain threats occur throughout the lifecycle from the design phase, through development, and into sustainment of the system. And while espionage has existed since the earliest conflicts, increasing dependence on information technology and systems has increased the threat from malicious insiders as well as their capability.

Moving operations into the cloud presents both threats and opportunities. Cloud services may provide significant cost and reliability enhancements; however, the physical loss of control over data presents a threat to operations. Carefully selecting appropriate missions and improvements in data protection schemes, including homographic encryption, could provide enhanced security as AF operations move to the cloud.

Opportunities exist in the development of mitigation measures to create vulnerability-free systems including creation of hardware and software roots of trust, the use of formal models for specification, and advanced automated verification tools. For example, the AF could leverage the methods used in DARPAs Crowd Sourced Formal Verification (CSFV) project for cost-effective analysis. Enhanced commercial and international partnerships will improve technology development, production lifecycles and secure supply chain management and enhance opportunities for a coordinated response.

5.3 Recommendations

Cyberspace root of trust: Develop trusted hardware, software, supply chain, out of band C2 and cloud services to improve security, agility, resilience and trust for AF networks and systems to achieve mission assurance in contested environments (OPRs: MAJCOMs, AFRL, AFLCMC, 24AF).

Integrated cyberspace operations: Develop offensive cyber capabilities to augment kinetic operations during wartime scenarios to affect strategic, operational and tactical missions. Develop persistent and/or dynamic access capabilities for collaborative missions across cyberspace, SIGINT, EW/EP, space, and communications to obtain a flexible full spectrum ISR capability in contested and A2/AD environments (OPR: ACC, AFSPACE, AFRISA, 24AF, AFRL).

Cyberspace situational awareness: Develop comprehensive cyber situational awareness capabilities for cyber superiority across blue and against red missions (OPR: AFSPACE, 24th AF, AFRL).

We refer the reader to the AF *Cyber Vision 2025* report for additional recommendations.

6. Global C2 and ISR

6.1 Trends, Threats and Opportunities

C2 and ISR are vital military capabilities, leveraged to confront an ever-increasing array of threats across all levels of war (strategic, operational, and tactical) from insurgents, to near-peer adversaries, who employ a wide range of capabilities up to and including weapons of mass destruction across all environments (permissive, contested, and highly contested). Global access to technology, worldwide connectivity, and increased access to all domains by our adversaries are closing the U.S. information superiority gap that will ultimately challenge our ability to dominate air, space, and cyberspace. The combination of increasing threats, information age advancements, and fiscal constraints simultaneously demand and enable the development of integrated, resilient, and innovative C2 and ISR "game changing" capabilities. To fully capitalize on these innovations will require new concepts of operations, and a new way of designing our force. Concepts of operation enabled by information-centric, interdependent, and functionally integrated organizations are the keys to future military success.

6.2 Game Changers

Figure 6.1 illustrates three C2 and ISR game changing themes with these recommendations:

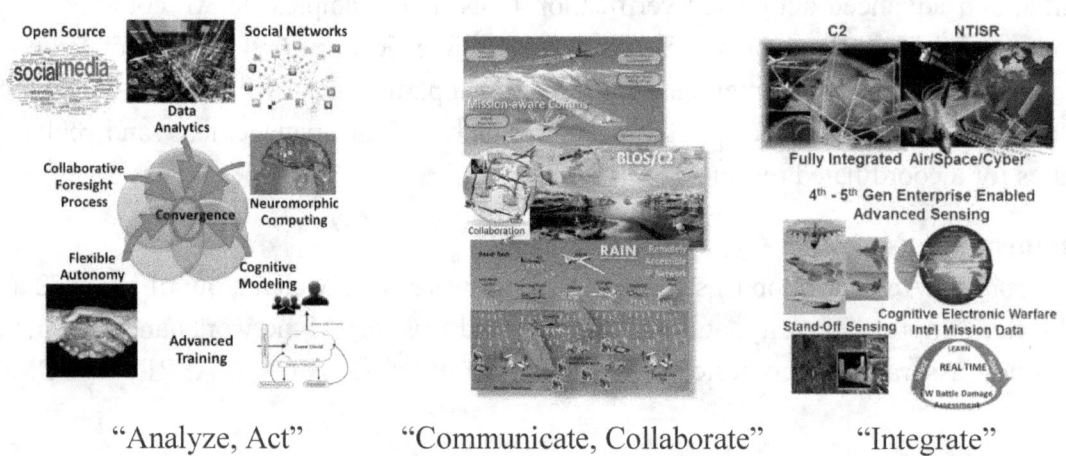

"Analyze, Act"   "Communicate, Collaborate"   "Integrate"

Figure 6.1: C2 and ISR Game Changing Themes

Innovative C2 and Analysis: *Ensure the speed of information exceeds the speed of engagement through automated analytics and planning.* Use data analytics, neuromorphic computing, cognitive modeling, and flexible autonomy to integrate platforms, sensors, and highly trained/educated operators for superior decision making. Develop flexible autonomy and all-source intelligence fusion and visualization technologies for enhanced analysis and planning capabilities for C2 and ISR.

Battlespace Networking: *Affordable, high-throughput air, space, and surface IP network providing real-time ISR and C2 collaboration.* Field a secure, self-forming, resilient, and agile IP network using existing infrastructure and advanced data link gateways enabled high capacity global C2 and tactical datalinks with mission-aware networking. Leverage this to support the build out of Joint Aerial Layer Network (JALN) using the JALN concept of operations and technology plan. Ensure C2 in a satellite-communications-denied environment, including support of the President of the United States mission essential tasks. Integrate coalition partnership capabilities through multi-level security (MLS) enabled networks.

Integration across Missions and Domains: *Shared understanding of the battlespace enabling anticipatory C2 and situation-aware tasking.* Fully integrate weapon systems and planning and direction, collection, processing and exploitation, analysis and production, and dissemination (PCPAD) across air, space, and cyberspace to achieve synchronized effects. Near-term opportunities include resilient space through small satellites; fully exploiting overhead persistent infrared data; enabling 5th generation aircraft to collect, process, and disseminate "targeted" ISR data; and automated decision aids for collaborative planning, dynamic execution, and assessment of operations enabled by a distributed resilient C2 and ISR enterprise. Fully integrate service and coalition forces by developing multilevel secure, message, and data formats.

6.3 Recommendations

- Develop flexible autonomy and all-source fusion technologies for enhanced analysis and planning capabilities for C2 and ISR. (OPRs: SAF/AQR, AFMC (AFRL & AFLCMC); OCRs: HAF/A2, NASIC, MAJCOMs)

- Field a secure, resilient, agile, and high capacity air-space-and-surface network to enable joint and multinational global C2 and ISR. (OPR: ACC A5/8/9, AFSPC; OCRs: SAF/CIO A6, HAF/A2, SAF/AQ, MAJCOMs, AFRL)

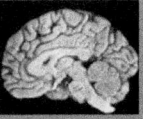 

Global Brain Science

Brain research promises break through knowledge in our understanding of the 86 billion neurons and trillions of connections that each of use daily for perception, cognition, and manipulation. Potential health, education, human, and computing benefits are game changing.

The DARPA SyNAPSE (Systems of Neuromorphic Adaptive Plastic Scalable Electronics) program aims to develop a computer with mammalian brain form and function housing 10 billion (10^{10}) neurons, 100 trillion (10^{14}) synapses, consuming one kilowatt, and requiring less than two liters of space. The program reported a cat-scale brain simulation (1.6 billion neurons and 8.87 trillion synapses) by IBM in 2009, a fully addressable 30 Gbits/cm^2 memristor array on top of a CMOS chip by HRL in 2011, the TrueNorth/Compass simulation of 530 billion neurons by IBM in 2012, and in 2013 expects IBM and Cornell University to report a second generation neurosynaptic processor with 1 million neurons per processor. Early demonstrations have included image and audio classification, spatio-temporal feature extraction, and robotic navigation. Recently, President Obama announced a $100M BRAIN (Brain Research through Advancing Innovative Neurotechnologies) Initiative across NIH, DARPA, and NSF.

Whereas the US had a dominant role in the Human Genome Initiative, brain research already enjoys strong global investment. There exist major brain research institutes in Asia including South Korea and Singapore. From 2002-2011 the European Union invested over €875M to support 187 brain research projects. In 2013, the Switzerland-based Human Brain Project, a collaboration of 80 European research institutes, won a ten year, €1.2 billion grant from the European Commission to build a human brain in a silicon substrate.

Potential benefits for leveraging this global R&D include not only revolutionary insights into traumatic brain injury and post-traumatic stress disorder but also augmented human cognition and very light weight, low-power machine autonomy across all Air Force core functions.

- Fully integrate weapon systems and PCPAD across air, space, and cyberspace to achieve synchronized effects. (OPRs: ACC, HAF/A2; OCRs: AFMC (AFRL & AFLCMC), SAF/CIO A6, SAF/AQ, MAJCOMs, HAF/A10)

7. Mission Support

7.1 Trends

Global annual expenditures in research are forecasted to reach $2 trillion by 2025 (See Figure 2.2) leading to breakthroughs in biology, materials, electronics, and software technology. However, key trends show that the integration of advanced technology into systems is an increasingly complex task. As just one example, the F-35 has over 25 million software lines of code along with new electronics and materials but the time required for development to initial operational capability (IOC) has grown to nearly 200 months, approximately three times as long as aircraft development time in the 1970s (See Figure 7.1). Increasing global research will also present the AF with increasing competition for high-quality scientists and engineers. By 2018, employment demand for technical talent is estimated to outpace degree production by over one million U.S. jobs.

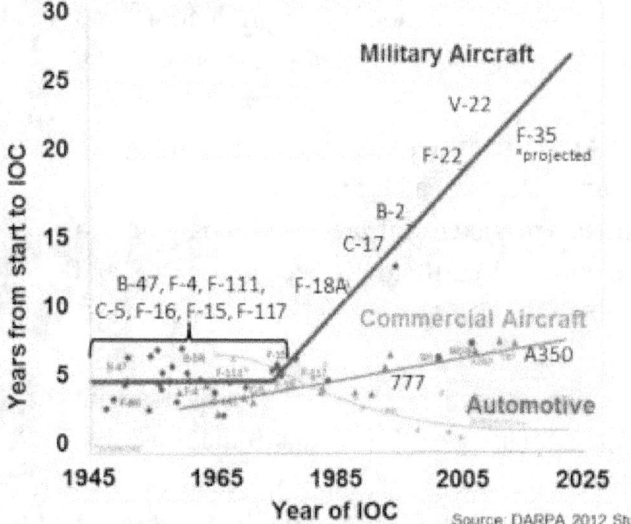

Figure 7.1: Increasing Complexity Slowing Development Cycle

7.2 Threats and Opportunities

Our inability to integrate and deliver innovation within acceptable timelines and our acquisition and life-cycle costs threaten to erode the decisive technology advantage that underpins today's AF capabilities. However, opportunities exist to exploit breakthroughs in new digital engineering tools that reduce the complexity of integrating advanced technology and shorten the development timeline. Coupling the new tools with a re-engineered prototyping process will enable more technology demonstrations than currently possible, provide earlier insight into technology maturity and suitability, and further lessen the development timeline. New modeling tools and demonstration processes alone will not solve the problem. A highly skilled and innovative technical workforce is also necessary. Opportunities in flexible management and "hands-on" engineering will enable us to recruit and retain the best talent and quickly respond to emerging technology challenges. Finally, we must work towards an Agile Combat Support structure that provides acquisition and operational speed, agility, and resilience. This will enable, for example, integrated air and missile defense to assure US forward presence and basing as highlighted in the AF Chief of Staff's Future Capabilities Game.

7.3 Game Changers

New System Design Tools – The Digital Threa:. Cross-domain, advanced physics-based modeling and simulation tools can reduce development cycle time by 25% through in-depth assessment of the feasibility and cost of integrating technologies into a system; provide data-rich assessment of cost and requirement trades; identify technology not ready for incorporation; quantify risk at critical decision points and avoid late defect discovery. Early system-of-system concept trades will enable optimized, disaggregated system architectures for interoperable environments; early digital design and manufacturing will enable agile development before metal is bent. The cross-domain, digital surrogate becomes the authoritative knowledge source managed across the system's life cycle. Coupling the "Digital Thread" (the use of digital tools and representations for design, evaluation, and life cycle management) with CONOPS and exercise environments will create rapid discovery and integration opportunities.

Re-energized Prototype Program: The AF can learn from industry, such as Scaled Composites' and SpaceX's successful demonstration of novel architectures and rapid prototyping processes to yield aerospace vehicles 50% faster compared to traditional acquisitions. Within DoD, prototyping has historically been pivotal for pre-acquisition risk reduction and concept validation, advancing new technologies, and workforce skills enhancement. The AF, with joint and industry partners, should re-energize prototyping efforts to provide early proofs of concepts and reduce technical uncertainty. An emphasis on technology demonstrations and open challenges will increase innovative breakthroughs, provide gap-filler capabilities, reduce risk aversion, and energize the workforce.

Expansion of Flexible Hiring Authorities: The laboratory personnel demonstration (Lab Demo) has proven successful in the recruitment and performance management of 2,500 scientist and engineering professionals. The same flexible hiring and employment authorities should be applied to the entire acquisition workforce to recruit technical employees with advanced degrees from a diverse pool of candidates 70% faster. In addition, offering dual technical and management career tracks combined with opportunities in rapid prototype environments will attract and retain top talent. This modern management structure ensures the AF has the talent to capitalize on game changing opportunities.

7.4 Recommendations
- Experiment with new cross-domain Digital Design Tools: (OPR: SAF/AQ)
  - Identify pilot programs to integrate System of System concept trades and digital design tools.

- Verify claims that the new tools reduce development time by at least 25% and save program costs.
- Reinvigorate a technology demonstration prototype program: (OPR: SAF/AQ)
  - Reallocate resources to increase the number of technology demonstrations.
  - Explore feasibility and utility of creating small, independent rapid prototype teams comprised of product centers, labs, users, academia, and industry.
  - Leverage external technical talent through "Open Challenges" and produce novel technologies and solutions at a fraction of the time and cost to conventional processes.
- Expand Laboratory Demonstration personnel program: (OPRs: HAF/A1, SAF/AQH)
  - Revamp AF personnel policies to grant Laboratory Demo authority to entire S&E workforce.
  - Enables rapid hiring (70% faster) of needed talent to quickly respond to technical challenges & opportunities.

8. Enabling Technology

8.1 Trends and Threats

Asymmetries in the value placed on human life, in the cost of military systems, and in historical dependence on technological superiority and its impact on adversaries drive the enabling technologies that should be developed by our nation over the next 15 years. We incur high costs to protect our warriors and care for them. Our weapons systems are expensive and potentially vulnerable to lower-cost countermeasures. Our technological superiority is no longer guaranteed given increasing global technological sophistication and productivity, as well as the global information grid. Its impact on cultures dissimilar to ours is poorly understood.

8.2 Opportunities

To respond to these asymmetries, we recommend targeted investments in the following five technology areas: (1) material sciences, (2) biotechnologies, (3) autonomous/robotic

---

Technology Focused Country Initiatives

The DoD basic research community has benefitted from leveraging long-term, high-value national level investments to advance specific capabilities in emerging technology areas. Heavy investment since the early 2000's into nano-technology, Korea ($2.5B), and nano-sciences, Taiwan ($1.2B), are providing tremendous collaboration opportunities with US government laboratory and university researchers.

US-Korea Nano-Bio-Info Technology (NBIT) Convergence Program

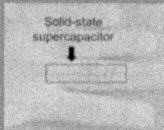

- New biscrolling technology has led to multifunctional carbon nanotube materials with enhanced properties.

Increased electrochemical performance, mechanical robustness & flexibility are key to emerging energy applications leading to new energy storage and power capabilities.

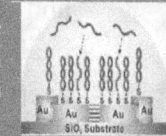

- High-throughput, facile nanostructure fabrication techniques have been developed & demonstrated.

These massively parallel nanostructure assemblies are enabling the advancement of real-time sensing and info technology.

US-Taiwan Nano-Science Program

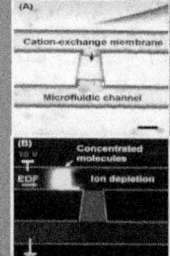

- Groundbreaking research in real-time concentration and detection of human performance biomarkers leading to devices capable of providing instant measurement of human readiness.

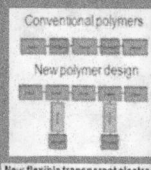

- Collaborative research is developing and engineering materials in order to produce next-gen flexible electrodes.

Uses range from robust flexible display screens to wearable electronic devices.

systems and platforms, (4) knowledge discovery and decision-making tools, and (5) social forecasting and influence. Advances in material sciences will lead to GPS-denied navigation capabilities that enable the maintenance of required navigational accuracies for hours instead of minutes. Advances in biotechnologies will bring about human-machine interfaces that can significantly reduce training and operations costs. Advances in autonomous/robotic systems bring decreased exposure of Airmen to harm, while simultaneously delivering desired effects with cheaper and more survivable weapons systems. Advances in accurate and timely knowledge extraction from enormous amounts of data will enable significantly better decisions to be made. Advances in social forecasting and influence will enable optimal use of military force to achieve national objectives, not just military objectives.

US R&D funding in 2012 was about 29% of the global R&D investment, down from 42% in 2000. Decreases in the U.S. percentage of global R&D investments are accelerating because of increasing global investments, especially in China and other Asian countries. For the AF to remain competitive, it must actively seek out and exploit investments in enabling technologies from global R&D, as discussed further in the Enabling Technologies Section (7) of the Appendix.

8.3 Game Changers

Cold-Atom-Based Navigation: Cold-Atom-Based Navigation to provide precisions many orders of magnitude greater than what can be achieved with the current laser-based navigation. Cold-atom-based navigation is currently at the applied research level, with initial systems anticipated to be available at a technology readiness level (TRL) of 6 by 2019 for large platforms (ships, large aircraft) (See Figure 8.1). Cold-atom clocks on a chip are necessary for smaller and less-stable platforms, such as ballistic missiles, satellites, and small unmanned vehicles. This technology is currently expected to be at TRL 6 by 2023. We also encourage development of alternative solutions such as vision-based navigation, chip-scale inertial navigation, and magnetic-field navigation technologies.

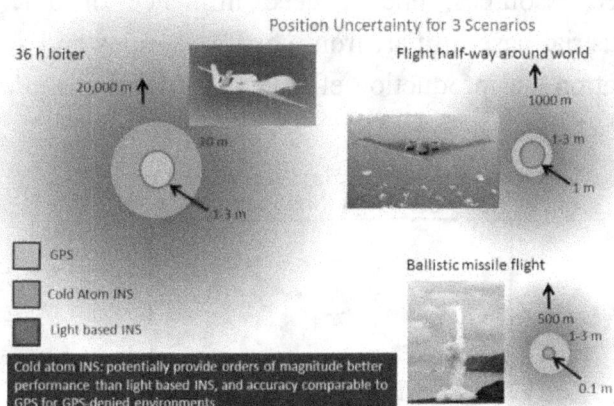

Figure 8.1: Position Uncertainty for Cold Atom Intertial Navigation System

Social Forecasting and Influence Tools: Incorporation of Social Forecasting and Influence Tools into policies, doctrine, and tactics. The near- and long-term socio-political aspects of military missions and weapons should be understood in the context of the human environment where they are used. For example, the Active Denial System, which is a directed energy millimeter wave beam weapon with a range significantly greater than any current non-lethal capabilities, might be utilized more effectively if the

psychological impact of this revolutionary weapon were more completely understood. Focused investment in this area is also critical to support other core AF needs such as indications and warning, cyber and strategic deterrence and global situational awareness. Such a capability would allow analysts to more completely assess current and future events, resulting in more informed and effective targeting and force-allocation decision making.

8.4 Recommendation

- Develop cold-atom inertial navigation system (INS) with a goal of a cold-atom INS on a large platform (e.g., transport, bomber) (OPR: AFRL; OCR: AMC, ACC, AFGSC)
    - AFRL Space Vehicles Directorate should engage AMC, ACC or AFGSC to draft a technology development roadmap for CFMP and establish regular technical exchange meetings with AMC/ACC/AFGSC by 2014.
    - By 2015 a tech insertion date should be agreed upon and mapped to the relevant MAJCOM POM. (AMC, ACC, or AFGSC)
    - Establish a Joint Capability Technology Demonstration (JCTD) timeframe by 2017 and a JCTD by 2019.

9. Manufacturing and Materials

9.1 Trends

Manufacturing employment in the U.S. is lower now than when the first personal computer was built in 1975, a consequence of globalization of manufacturing and technology proliferation (See Figure 9.1). For the DoD and AF, this situation is aggravated by a limited trained US domestic workforce, reduced resources, and reduced influence of defense materials and processing needs on the industrial base, in part from small quantities and sporadic acquisition. However, because of automation and production efficiencies, global manufacturing output has risen over time (See Figure 9.2).

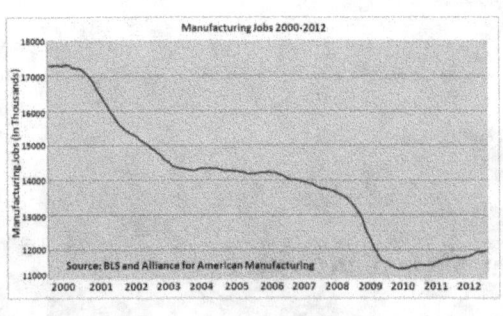

Figure 9.1: Manufacturing Jobs Decreasing

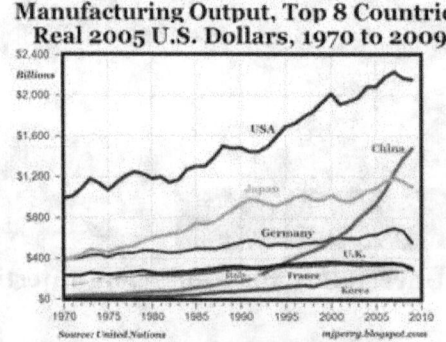

Figure 9.2: Manufacturing Output Increasing

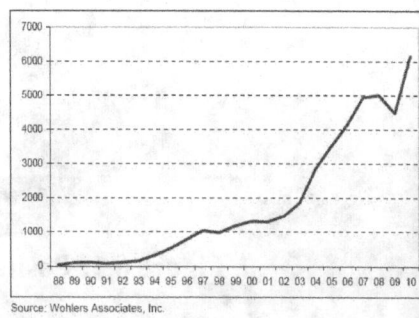

Figure 9.3: Additive Manufactur[ing] Machine Sales Increasing

9.2 Threats and Opportunities

The erosion of the manufacturing base bodes ominously for the ability of the AF to design, develop, manufacture, and deploy trusted advanced technologies on a time scale consistent with the emergence of new threats. Global trends toward more agile and distributed manufacturing will only exacerbate the challenges, especially as regards trusted sourcing. Speed to application and deployment is critical to maintaining the technological advantages of the AF.

9.3 Game Changers

Exploiting the three game-changing opportunities below will help the AF meet the need for more rapid development and deployment. The recommendations represent the first steps on the path to future game-changers.

Advanced Manufacturing: Advanced manufacturing technologies including additive (See Figure 9.3), 3-D, and direct digital printing, will enable open architectures that permit rapid prototyping, mission specific reconfigurability; material tailoring for specific applications (See sidebar); efficient small lot productions; better systems, faster and cheaper. Advanced manufacturing technologies will deliver products when and where needed and will facilitate multi-functionality, with manufacturing cycle time improvements from 60% in design phase to 30% in automated assembly. On-site Advanced manufacturing could allow for instant part replacement for battle damage repair.

Redefined Qualification and Certification Paradigm: Redefining the Qualification and Certification Paradigm will allow rapid utilization of products from advanced manufacturing (efficiently from prototype to practice). The new paradigm will eliminate the excessive

Global Materials Science

Novel materials portend revolutionary benefits in strength, weight, agility, and electromagnetics with promising range, stealth, and survivability game changing effects. As overseas investments in materials and manufacturing grows, it will be imperative to maintain expert global engagement of science and technology investments relevant to Air Force core functions.

For example, in 2004 at Manchester University in the United Kingdom, researchers first separated a single-atom layer of graphene by peeling away layers of pencil lead using Scotch tape yielding a material that was 100-300 times stronger than steel, more conductive than copper, impermeable to gases, and had unique optical properties. In 2010 these partially AFOSR-funded researchers were awarded the Nobel prize for revolutionizing electronics to be lighter, stronger, more flexible, and faster.

In 2013, with the ambition of becoming "Graphene Valley" (like "Silicon Valley"), the European Commission announced a €1 billion graphene initiative involving 126 academic and industrial groups from 17 European countries. Promising lighter cars and airplanes, carbon fiber is becoming a major element in Airbus doors, the Eurocopter's airframe, and the mass produced urban electric BMW i3, which will have most of its chassis and body made of carbon-fiber reinforced plastic.

Why would the Europeans make such big bet on R&D investment? The graph below illustrates the correlation between the level of R&D investment as a % of GDP over a five year period and the resulting GDP growth.

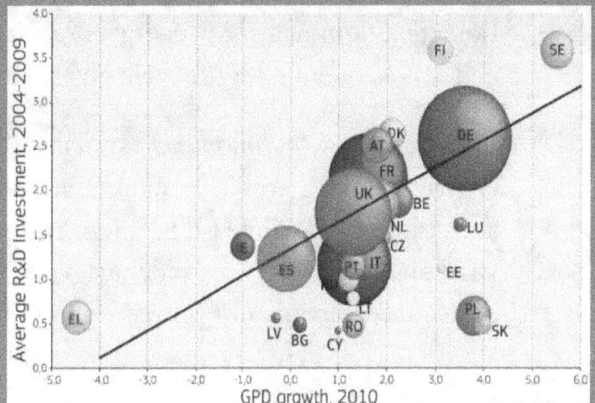

development times for complex capability systems (15-20 years) by inclusion of concepts such as defined and finite system life, qualification and certification as "adequate" for this application for this length of time, and process qualification and certification *vice* component qualification and certification.

Digital Thread and Digital Twin: The concept of a digital thread/digital twin comprised of advanced modeling and simulation tools that link materials-design-processing-manufacturing (Digital Thread) will be the game-changer that provides the agility and tailorability needed for rapid development and deployment, while also reducing risk. State Awareness and System Prognosis advantages will be achieved through the Digital Twin, a virtual representation of the system as an integrated system of data, models, and analysis tools applied over the entire life cycle on a tail-number unique and operator–by-name basis. M&S tools will optimize manufacturability, inspectability, and sustainability from the outset. Data captured from legacy and future systems will provide the basis for refined models that enable component and system-level prognostics. Archived digital descriptions of new systems would greatly facilitate any subsequent re-engineering required in the future. Human performance monitoring will enable adaptation of systems to the "mission capable" state of the operator.

9.4 Recommendations

- To increase life cycle affordability and rapid development, define pilot programs to instantiate the Digital Thread/Twin from concept development though disposal
  (OPR: AEDC/CZ)
- To more rapidly provide the AF with the advantages of the latest materials & manufacturing advances, establish a working group to: (OPR: AFRL/RX)
  - Identify and eliminate obstacles that limit AF exploitation of the benefits of additive and other agile manufacturing methods
    (OPR: SAF/AQ, AFMC/EN/A2/5, AFLCMC/EZP)
  - Identify AF specific requirements and research needed to enable agile manufacturing to meet them
    (OPR: AFLCMC/XZI, SAF/AQR, AFMC/A2/5)

10. Logistics and Transportation

10.1 Trends
Logistics dominates AF energy use, drives mobility requirements, enables/hinders operations, and drives overall AF lifecycle costs. Key trends include:

Robotics and autonomous systems: Autonomous vehicles capable of operating in any environment in which humans currently drive or pilot, learning, and adapting to changing scenarios are predicted during the next 15 years. Current automated ports, like the wharves in Brisbane, Australia (see sidebar), saw a 27% reduction in labor, 70% savings in maintenance costs, and dramatic (18 fold) drop in injury rates.

Energy: Although the volume of air traffic will double by 2027, the commensurate energy bill increase should be slightly less (1.5x), resulting in part from more energy efficient platforms and operations. Diversity of fuel sources should increase (e.g., alternative fuels and recovered petroleum). While the U.S. may regain its status as a net exporter of oil, most transportation industries predict a rise (50%) in fuel costs by 2030. Exacerbating this, energy supply chains are enticing targets.

10.2 Threats and Opportunities

Logistics breeds logistics - current operations have inefficiencies and large tails. For example, a case-study in the "fully burdened cost of logistics" (moving and sustaining two JSTARS aircraft) suggests the requirement for large-scale airlift, tanker, fighter, and combat support, results in a forward footprint of over 1,000 people (deployed and on-site) and recurrent energy costs of over $25M/month to secure, operate and sustain operations (See Figure 10.1).

Operations efficiency and cyber threats need to be addressed early in the acquisition process. Advanced materials, new propulsion systems, and aerodynamic improvements could result in significant (20-40%) reductions in logistic requirements and fuel operating costs by 2030. Quantum validation and verification of software and use of trusted foundries would contribute to better cyber security.

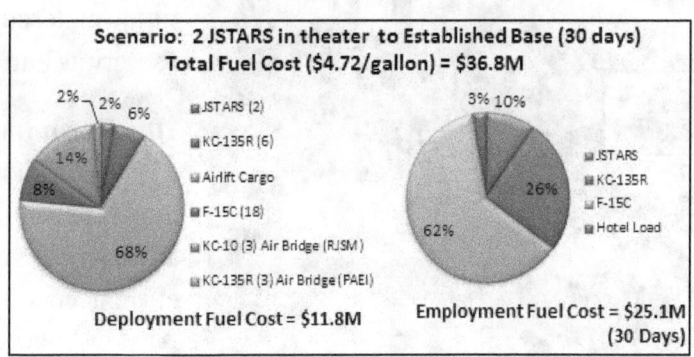

Figure 10.1: Fully Burdened Cost of Logistics

10.3 Game Changers

Several technological advances hold promise to reduce the logistic requirements of the AF:

Autonomous/Remotely Operated Systems: Home station logistics operations and delivery will be enhanced with increased use of robotic or remotely operated systems. Deploying these systems should reduce the forward footprint. Material processing and handling (armaments and cargo), servicing, maintenance, emergency response, protection, and base surveillance are all potential automation/remote operation targets.

On-Site Production: Advances in manufacturing technology like "3-D printing" would allow rapid generation of needed devices and parts. Use of indigenous resources and assets, including recycled materials, offer flexible and potentially cost-saving procurement options.

Most Automated Port in the World

Agility and resilience in logistics is a hallmark of successful militaries. But there is much to learn from commercial operators. Now operating a third generation port with the world's largest autonomous robots, Brisbane Port is a global shipping center that employs 27 autonomous, large mobile Autostrads with 2 cm precision location using millimeter wave radar and model based control to handle 800k Twenty-foot Equivalent containers yearly.

Port Brisbane, Australia

Single Operator C2

Model Based Control

Facing a highly competitive market ($22 an hour wage in Australia versus $3/hour in China), Patrick Port Logistics achieved competitive advantage through automation by leveraging technology originating from science at the Centre for Field Robotics at Sydney University and supported by NICTA (National ICT Australia). Today, the port enjoys a 27% reduction in labor, 40% lower fuel costs, 66% increase in logistics velocity, and 70% reduction in maintenance. Automation was only a 10% premium on port construction cost so payback occurred in less than 2 years and labor to revenue was reduced from 50% to 21%. Moreover, injuries were reduced 94% as shown in the chart below.

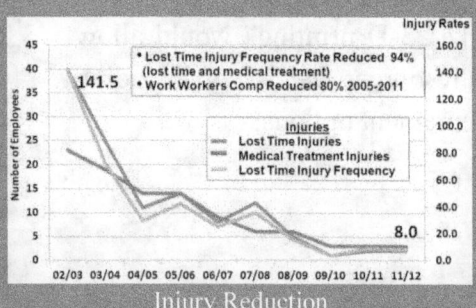

Injury Reduction

Improved Efficiency: Transportation of energy and equipment for power production consumes much of the supply chain capacity, better energy efficiency has a compounding benefit. Advances in computer processing and algorithms will provide more optimal routing, scheduling, and tracking. Adopting commercial best practices will improve in-transit visibility and customer confidence.

Precise, Direct Delivery: Eliminating intermediate nodes generates direct delivery with compounded reduction in logistics requirements. Advanced precision airdrop, RPAs and autonomous and robotic technologies could be employed to provide just-in-time materials to austere locations reducing base footprints and storage requirements.

10.4 Recommendation

The AF should conduct a series of field tests, experiments, and challenges to reduce the logistics and combat support footprint of the AF by 50% (over current costs) by 2025. (OPR: AFMC, OCRs: ACC, AMC, AFGSC, AFSOC, HAF/A4/7) A stakeholder IPT should:

- Define and validate a methodology to measure "fully-burdened cost of logistics" against current baselines.
- Select representative technologies (e.g., autonomous warehouse robots, remote-sited 3-D printing capability, secure supply chain sourcing) and mission functions for consideration, test, and evaluation.
- Generate integrated roadmap for development and test, and conduct exercises to verify savings and improvements in operational capability as a result of logistics footprints.

11. Energy

11.1 Trends

Energy is critical to every AF mission (See Figure 11.1). Access to sufficient energy is essential to assuring air, space, and cyberspace missions; however global industrialization is increasing energy demand and global political volatility is negatively impacting energy supplies and cost. World energy consumption is forecast to grow 30% from 553 quadrillion Btu in 2013 to 721 quadrillion Btu in 2030. Global oil prices, which currently hover around $100 per barrel, are projected to rise to between $130 and $200 by 2030, but the true upwards potential is unbounded. Recent U.S. shale gas production increases may moderate, but not reverse, these trends.

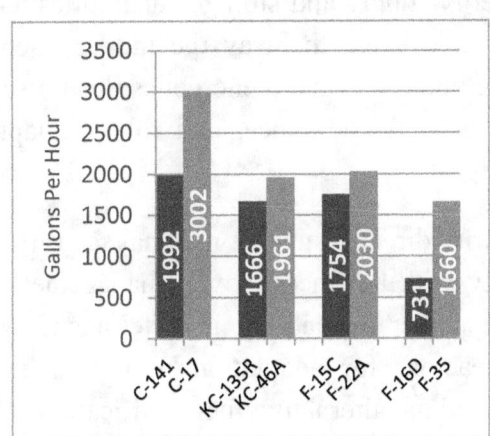

Figure 11.1: More Capability, More Fuel

11.2 Threats

Energy dependence is a strategic risk. Energy is increasingly targeted by adversaries as a center of gravity. Future aviation fuel and electrical grid supply disruptions could drive correlated mission capability reductions. Increases in energy costs result in must-pay bills which siphon money and slow or degrade sustainment and acquisition projects. In FY12 alone the AF reprogrammed

High Energy Solid State Lasers

High energy solid state lasers require gain material dimensions that are not currently available through single crystal growth methods. As a result, global researchers have delved into transparent laser ceramics, as one possibility, to satisfy the requirements of larger, more powerful solid state lasers.

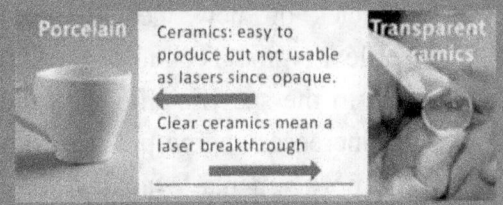

AFOSR has led the DoD in collaborations with world-leading labs in Japan developing these transparent laser ceramics. Power densities achievable with transparent ceramics have improved from kilowatt-class materials in 2005 when AFOSR initiated collaboration with these labs to megawatt-class materials in 2012.

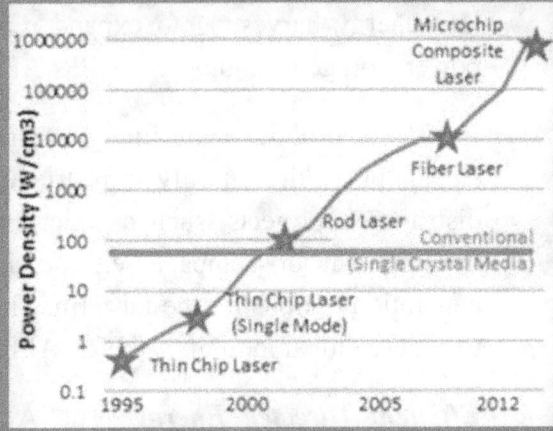

Advanced processing techniques developed in Japan are taking specialized ceramics, similar to porcelain, and fabricating transparent composite laser waveguides capable of high power densities with excellent thermal characteristics. These materials are enabling the next generation of solid-state laser systems. This engagement with Japan in laser gain material fundamental research paves the path for enabling laser development for future military applications.

$500M from key weapon systems programs to pay higher fuel costs.

11.3 Game Changers

With negative trends impacting energy availability and affordability, the AF must leverage the *Energy Horizons* (2011) recommendations and pursue technological innovation to reduce demand, increase supply, and improve resiliency.

Advanced Propulsion: Jet fuel accounts for 86% of AF energy use, thus the AF should reduce demand by focusing technology investments to advance propulsion and aerodynamic efficiencies on new weapon systems. AFRL should continue development of ADaptive Versatile ENgine Technology (ADVENT), which reduces fuel burn by matching the engine airflow to the specific flight envelope, Highly Efficient Embedded Turbine Engine (HEETE), which increases engine pressure ratios, and complementary thermal management and adaptive cycle improvements to achieve 25-35% fuel burn reductions. AFRL should also continue to pursue aerodynamic improvements for future aircraft such as laminar flow optimization, blended wing, and lifting body construction to deliver 15-25% energy efficiency improvement. (OPRs: AFRL, SAF/AQR)

Energy Storage Density: The AF should pursue energy supply and storage capabilities. Order of magnitude advances in energy storage density can change the way the military deploys energy in air and space. The AFRL should lead development of nano energetics, lead research into energy harvesting to expand remotely piloted aircraft endurance, and follow adaptable power storage technologies. (OPR: AFRL)

Resiliency and Security: Energy resiliency and security directly support national security. The AF should address utility infrastructure vulnerability with advanced power management and distribution projects, such as microgrids, to ensure power availability near term and watch development of compact self-contained nuclear reactors for adaptation into base energy generation options for the far term. The AF should continue alternative fuel certification efforts to ensure global mobility. (OPRs: AFRL, AFLCMC)

Efficient Directed Energy: The AFRL should continue development of directed energy technology as it could enable efficiency enhancements as well as revolutionary capabilities. Beaming power may enable currently impractical energy intensive applications, such as certain space based capabilities. Replacing kinetic anti-missile weapons reduces and accelerates the logistical tail of replacement missiles, as directed energy weapons recharge and fire effectively with the equivalent of 1-2 gallons of fuel. (OPR: AFRL)

11.4 Recommendation

In summary, the AF should institutionalize energy consumption considerations across the requirements and acquisition continuum. (OPR: SAF/IE, SAF/AQ, Energy Council)

12. Communications, Information Technology and Financial Services

12.1 Trends

Real-time and pervasive information technology and communications are critical drivers to enable a competitive advantage in military and global sectors. Across consumer services from Amazon to Wall Street, communication and processing time is measured in milliseconds and influences billions of dollars in revenue daily. Government and industry are experiencing increasing bandwidth requirements in a communications spectrum with decreasing availability. The need to handle big data and act faster, combined with an exponential growth in computing density, has led to the increasing adoption of automation and less dependency on manual labor. Information technology and communication, whether in the form of consumer electronics or AF assets, is largely a widely available global resource irrespective of demographics. With technology refresh occurring approximately every two years, proper technology governance models and intellectual property rights are essential to maintain a competitive advantage.

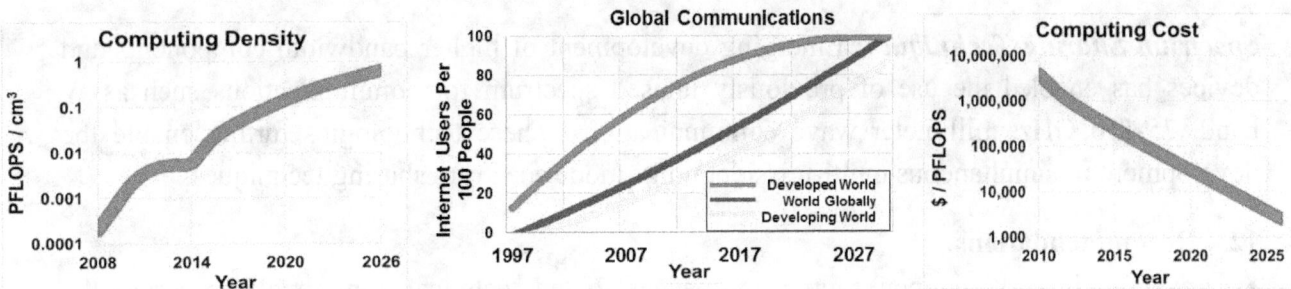

Figure 12.1: Computing Performance, Cost and Global Communications Trends

12.2 Threats and Opportunities

Big data represents both a threat and an opportunity. The increasing financial sector demand to handle massive data collection, storage, processing, and analysis is comparable to emerging AF capabilities in C2 and ISR. Real-time analytics threaten to be overwhelmed by data volume, velocity, variety and trust. The big data era presents opportunities for innovative technologies where traditional data analysis methods are overwhelmed. Opportunities for advancements can be realized by developing affordable, secure and intelligent computing architectures for massive analytics.

Manufacturing technology used in computer chip fabrication is reaching its physical limits in terms of area, performance and power. Novel and revolutionary solutions to achieve greater computing capacity are emerging in the areas of high density system integration, nanotechnology and quantum electronics.

De-centralized connectivity which utilizes a distributed backbone is an attractive topology for both AF and business enterprises as it allows for easy expansion and limited capital outlay for growth. Opportunities in this area include scalable mobile ad hoc networks, near-field communications, software-defined radios, and technologies that enable persistent and pervasive communication links.

12.3 Game Changers

Symbolic Inference Models and Neuromorphic Computing: Large-scale symbolic inference models and neuromorphic computing architectures are essential technologies to achieve affordable (100X reduction in computing cost), agile, cognitive, and trusted systems. These technologies are capable of ingesting and processing massive data (100X increase in data analytic performance).

3-D Chip-Stacking: 3-D chip-stacking technology will provide over 100X increase in computing density and energy efficiency over the next 15 years, driving petascale computing in embedded systems. Additionally, advances in multifunctional nanoelectronics and nanomaterials for low-cost and sustainable energy can provide another 100X improvement in size, computing performance, and power efficiency over the next 15 to 25 years."

Spectrum Sharing Techniques: Emerging development of higher bandwidth components and devices has enabled the use of previously unused spectrum for communications such as W band, 75-110 GHz, millimeter wave communications. These technologies further enable the development of simultaneous multi-mission, multi-mode spectrum sharing techniques.

12.4 Recommendations

- Leverage innovative open source approaches to tap technical commonalities across the Communication, IT, and Finance sectors, including co-investing with international partners. (OPRs: SAF/AQ, AFRL; OCR: SAF/IA)
- Lead S&T for high performance embedded computing across air, space and cyber A2/AD environments (OPR: AFRL) for size, weight, and power constrained applications exploiting advances in 3D chip stacking, nano-technology, and quantum computing.
- Develop open architecture post-JTRS "cognitive" communications for agile, networked, cost effective communications in A2/AD scenarios.
  (OPRs: AFRL, AFLCMC)
- Leverage and adapt global sector expertise in "big data" analytics across multiple disparate sources: (OPRs: AFISRA, AFRL)
  - Develop real-time analytics for ISR (Cyber/SIGINT/EW) akin to financial sector.
  - Focus petascale computing on neuromorphic and symbolic approaches to computational intelligence.
  - Adapt discovery/fusion ideas from IT/Finance to multi-int ISR problems.

13. Pharmaceutical and Health Care

13.1 Trends

The global cost growth of the pharmaceutical and health care sector is unsustainable (See Figure 13.1). This growth is accelerated by the cost and time to bring a new drug to market, the ineffectiveness of breakthrough and block buster drugs, and the rise in age-related and preventable chronic diseases, underscoring the need to revolutionize the sector. Fortuitously, the global proliferation of mobile, sensing and data technology (e.g., Figure 13.2) has set up the necessary infrastructure for four critical technology drivers that are transforming the sector, including: 1) mobile health and the quantified self, 2) nanomedicine, 3) genomic sequencing and 'omics,' and 4) big 'my' data (see Pharmaceutical and Health Care Section (13) of the Appendix for details).

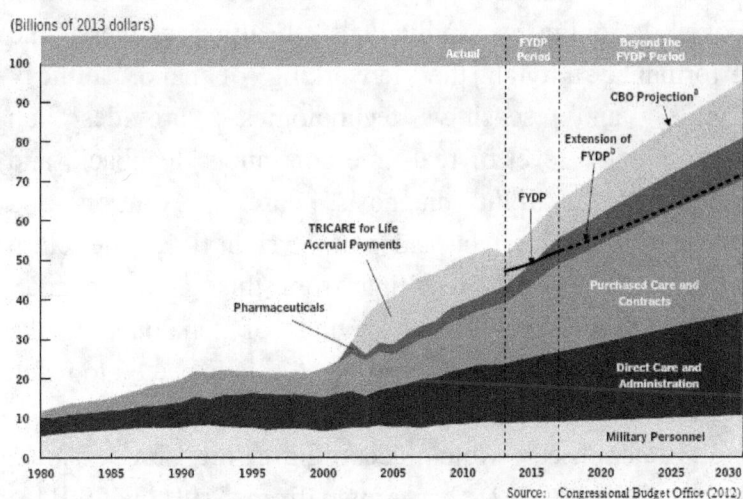

Figure 13.1: Rising DoD Health Cost Plans

13.2 Threats and Opportunities

Given the key trends, potential threats fall into two major categories: privacy/security in the near-term and biological weapons in the far-term. For privacy and security, threats include malicious biohacking and external control and manipulation by adversaries, genetic and medical identity theft, and an increasing difficulty to keep secrets and avoid detection. For biological weapons, threats include a new class of intelligent, precise bio-terror weaponry. The trends also reveal new opportunities in health care, performance, and selection for the AF. For health care, opportunities center on seamless care from monitoring, to diagnosis and treatment, to therapy, from the first responder to the in-garrison medical team to the caregiver at home. Opportunities for performance augmentation include continuous performance feedback for self-improvement and individualized training regimens in the near- and mid-terms, respectively, and optimized human-machine teaming in the far-term. Opportunities afforded by the 'omics' and "Big 'My' Data" show potential for empirical selection and matching of the right person for the mission.

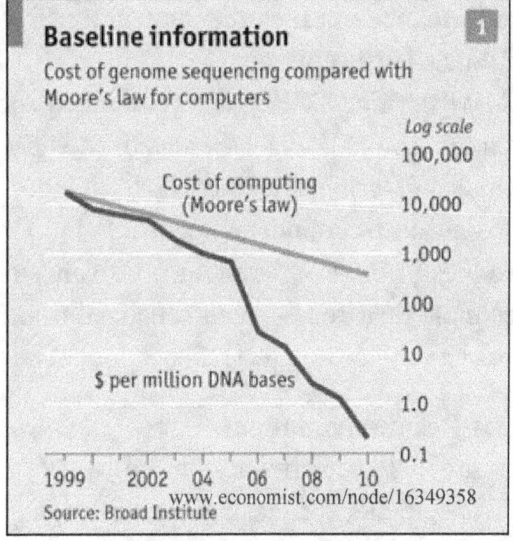

Figure 13.2: Dropping Genomics Cost

13.3 Game Changer

Personalized Health and Performance: The game changer for the AF is personalized health and performance. It is the result of the superconvergence of the trends identified above. It optimizes individuals' health, wellness and performance through the networking of nano, 'omics,' mobile, and sensing technologies, provides an unprecedented level of real-time continuous feedback, and results in the right diagnosis, care, prevention and intervention for the right person at the right time. The return on investment (ROI) resulting from this game changer, derived from estimated cost savings of personalizing the Pharmaceutical and Health Care Sector, is expected to reach into the billions. It confronts the unsustainable cost of military health care which, according to the Congressional Budget Office (2012), has grown from $19B in 2001 to $53B in 2012, and is expected to escalate to $95B in 2030 (See Figure 13.1). Moreover, personalized health and performance expands the strategic vision postulated by the Military Health System (MHS) and the AF Medical Service (AFMS). The MHS has embraced personalization via genomics and Patient-Centered Medical Homes, both of which are critical to the game changer, but fall short of closing the continuous feedback loop in the health and wellness control system. Advancements in mobile technologies are key to personalized health and performance and serve to close this feedback loop.

13.4 Recommendations

In order to determine the utility of mobile technology advancements for closing the health and wellness loop for the AF, the following recommendations are proposed:

- Given a potential savings of over $200M annually with personalized medicine, analyze the ROI for type 2 diabetes control via personalized health and wellness apps and self-tracking devices (OPR: HAF/A9) and conduct a demonstration program to validate the ROI specific to TRICARE-enrolled beneficiaries. (OPR: HAF/SG)

Global Mobility Innovations from International Investments

Leveraging capabilities from the *Istituto Nazionale per le Malattie Infettive* in Rome, the World Health Organization (WHO), and unique DoD and civilian expertise, the Air Force Air Mobility Command recently deployed the Highly Infectious Patient Isolation Transport Unit. This capability enables the safe transport and management of stabilized biologically contagious patients and is FDA-approved, airworthiness certified, and NATO litter compatible.

And leveraging German capabilities, our medical personnel perform extracorporeal membrane oxygenation using a heart-lung bypass device, simultaneously providing cardiac and respiratory support to treat and transport severely wounded combat casualties. The device circulates and oxygenates blood (filters Carbon Dioxide from the bloodstream, inserts Oxygen directly into arterial blood) providing diseased or battle damaged lungs an opportunity to heal. These efforts not only avoided significant development cost, but also saved years of development time and afforded rapid fielding of improved capabilities to life-saving aeromedical evacuation

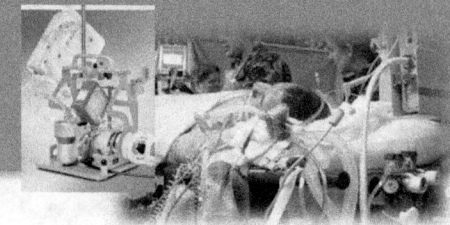

- For performance optimization, conduct a pilot project to empirically determine the ROI for self-selected fitness and health management applications and biomonitoring devices used by the AF Special Tactics Teams. (OPR: HAF/SG; OCR: AFSOC)

14. Education and Training

14.1 Trends

Education and training is facing a perfect storm: increased costs (See Figure 14.1), constrained resources, increased operations tempo, and unprecedented complexity of requirements for airmen. Concurrently, improvements in information technology enable virtual delivery of training and education, to include massive open online courses (MOOC, a Stanford MOOC logged over 150,000 students), instructional use of games and simulations (the $78B gaming industry projects growth of 10% per year), and an explosion in social media (Facebook reports over 1.5 billion users in 70 languages). The National Science Foundation (2010) reports that U.S. STEM graduation rates lag behind nations like China and the European Community (See Figure 14.1), and motivation for scholarly activity and research is stifled and undervalued as business and industry emphasize production over investment.

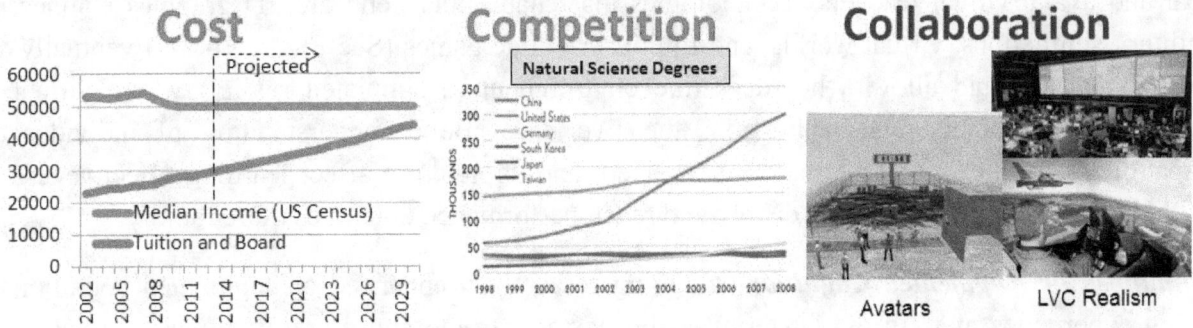

Figure 14.1: Cost, Global Competition and Virtual Collaborative Training Opportunity

14.2 Threats and Opportunities

Budget stresses continue to impact education and training. Nonetheless, the opportunities for the AF are exciting. Realistic, adaptive and interactive scenario-based education and training can provide integration of real-world lessons-learned adapting to warfighter's individual needs. As improvements continue in adaptive and intelligent web-based systems, mobile networks, desktop trainers, wearable devices, visualization, virtual spaces and avatars, the AF can leverage integrated, personalized learning that allow seamless, relevant, mission-focused simulations and courses to be available when needed, though at non-trivial programming and information security cost. Better designed and validated tools and metrics will optimize training outcomes and enable auto-capture measures of performance and effectiveness. Predictive testing in recruiting, selection and utilization will better match personal talents and dispositions with developmental opportunities to better pinpoint the right person for right job. Finally, we can build upon effective conventional approaches such as internships and mentorships (e.g., recent information assurance internships with international partners such as the Royal Air Force).

14.3 Game Changers

State-of-the-Art Information Technology Communication Backbone: Improvements in visualization technology, expert systems, natural language processing, social media, immersive environments and communications/networking capabilities (bandwidth and coverage) will help make the walls of traditional brick and mortar classrooms transparent, and possibly eliminate them altogether. An AF-wide state-of-the-art information technology communication backbone

Figure 14.2: Virtual Aviation

will provide secure and unrestricted bandwidth/connectivity to facilitate full-spectrum use of virtual learning technologies. Virtual classrooms will be the norm, allowing live-virtual-constructive environments where students learn on demand by interacting with combinations of real and avatar/virtual teachers. Technologies that enable this trend are 3D graphics engines, military simulations, virtual worlds, and multiuser online games (See Figure 14.2). Eventually, virtual reality could allow a holodeck-like environment (a simulated reality) where virtual participants are indistinguishable from the live ones. Better understanding of individual capabilities and how factors such as diet, sleep, and stimulation affect learning effectiveness will allow students to learn more deeply and reach performance outcomes more rapidly.

Enhanced or Augmented Cognition: Ongoing research in enhanced or augmented cognition using cybernetics may, in the far term, enable faster learning and more effective information and skill-set recall. This should accelerate Airmen's ability to acquire skills in the AF Institutional Competency List and produce not only better shooters but also better critical thinkers who are creative and innovative in their approaches to increasingly complex future environments.

14.4 Recommendations

- Pursue a live, virtual, and constructive training and education initiative:
  (OPR: HAF/A3; OCRs: ACC/A3, AETC/A3 (plus AFSOC & AMC), AFRL/RHA)
  - Efficiently mix live and virtual with the increasingly realistic constructive players, software agents and job aids.
  - Devise persistent metrics/assessments in achieving readiness goals.
  - Expand scope: include strategic/operational level warfare, more players, and international cooperation.
- Support STEM-producing advanced degree education programs.
  (OPR: HAF/A1; OCR: AETC, AFA, AFRL/AFOSR)

15. Conclusion and Summary

Global Horizons is an S&T vision and blueprint to help the *Air Force* achieve the "assured global advantage" across core AF functions. *Global Horizons* recognizes that all our core functions depend on global domains and that our warfighting mission systems are both threatened and enabled by

> *"A time of unprecedented shifts in the world order, new global challenges, and deep global fiscal uncertainty"*
>
> **Honorable Chuck Hagel,**
> **Secretary of Defense**
> **3 April 2013**

global industrial sectors. Furthermore, these global domains are increasingly contested and/or denied from increasingly capable adversaries. Our current environment is characterized by constrained resources (e.g., financial, human, time) derived from federal deficits, limited production of U.S. STEM graduates, and increasing threats in the commons. Yet global industrial sectors present important opportunities.

Summary key findings of *Global Horizons* include:

- Constraints (natural resource, human, budget, time) compel efficiency, speed, and focus in RDT&E and operations.

- S&T recommendations from the 2010 *Technology Horizons* (autonomy, human effectiveness), 2011 *Energy Horizons* (generation, use, distribution), and 2012 *Cyber Vision 2025* (mission assurance, resiliency and agility, human machine integration, trust) remain valid and are consistent with *Global Horizons* but require sustained focus.

- Global domains will be increasingly contested, congested, and competitive, adversely impacting AF core functions.

- Strategic opportunity exists to leverage the $1.4 trillion annual R&D investment in global industrial sectors.

- Rapid and economical leverage of global invention and innovation will be essential to sustaining our advantage.

- Supply of educated talent will be constrained and contested.

Global threats and opportunities described in the sections above are detailed in an accompanying Appendix which includes near, mid and long term technology roadmaps where the AF should lead, follow, and watch. They compel action across the AF enterprise on the following key study recommendations:

- Enhance global S&T vigilance to anticipate and counter strategic threats. (OPRs: NASIC, AFRL/AFOSR, AF/A2, AFISRA)

- Focus AF S&T on game changers with associated revolutionary concept of operations (CONOPS) (OPRs: AFRL, MAJCOMs) in these rank-ordered, priority areas:
  - Trusted and resilient cyberspace[3], assured PNT (e.g., cold atoms, vision-based navigation), hypersonics and directed energy weapons, bio-inspired computation, advanced materials and manufacturing, personalized health/performance
- Employ agile and innovative acquisition approaches (e.g., grand challenges/prizes, crowdsourcing, Advanced Concept Technology Demonstrations, prototyping); Foster partnerships (e.g., DARPA, NASA, DoE); Shape doctrine, policy, and processes (RDT&E, digital thread) for agility, speed, and economy and regularly review and update to take advantage of S&T and capability development (OPR: SAF/AQ, AFMC, AFRL)
- Proactively track and leverage AF relevant global industrial investments (e.g., transportation, manufacturing, health care) and pursue strategic international partnering. (OPRs: SAF/IA, AFMC, AFRL/AFOSR, MAJCOMs)
- Inspire and focus accession, development and retention of STEM workforce. (OPRs: AF/A1, SAF/AQ, AETC, AFA)

Air Force leaders at all levels should make global advantage a priority by taking concrete actions in their own units. Realizing the full promise of *Global Horizons* will require concerted and sustained AF leadership and external partnership to ensure the necessary cultural change and organizational evolution to sustain assured global advantage. In addition, since no plan survives contact with the future and with rapid technology progress, *Global Horizons* should be revisited at least every 5 years to update the AF S&T blueprint.

In conclusion, our sustained global advantage relies upon our ability to assure access to global domains and leverage the global industrial centers of gravity to ensure victory in future major military conflict. *Global Horizons* provides a critical element of our path to success in peacetime, during humanitarian and disaster relief, or in military conflict. Working as a team, in full partnership with international partners, other services, agencies, national laboratories, FFRDCs, industry and academia, the AF must strategically leverage global opportunities in the global industrial sectors to deter threats across air, space, cyber, C2, ISR and mission support to ensure its future ability to fly, flight, and win in air, space, and cyberspace.

---

[3] Joint Pub 3-12 defines cyberspace as "A domain characterized by the use of electronics and the electromagnetic spectrum to store, modify and exchange data via networked systems and associated physical infrastructures." This includes BLOS/C2 and joint and coalition airborne networks.

16. References

2013 Global R&D Funding Forecast. December 2012. Battelle. www.rdmag.com

AF Doctrine Document (AFDD) 1, *Air Force Basic Doctrine, Organization, and Command,* 14 October 2011.

Air Force Doctrine Document (AFDD) 2-0, *Global Integrated Intelligence, Surveillance, and Reconnaissance Operations.* 6 January 2012.

Air Force Doctrine Document (AFDD) 3-12, *Cyberspace Operations*, 15 July 2010, incorporating changes 30 Nov 2011.

Air Force Doctrine Document (AFDD) 3-13, *Information Operations*, 11 January 2005, iincorporating changes 28 July 2011.

Air Force Chief of Staff's Future Capabilities Game 2013 Quick Look Report. 10-15 March 2013. LeMay Center Wargaming Institute, Maxwell AFB, AL.

Air Force Global Partnership Strategy, 2011.

Allan, P.; Osborn, E.; Bloom, B.; Wanek, S. and Cannon, J. August 2011. The Introduction of Extracorporeal Membrane Oxygenation to Aeromedical Evacuation. *Military Medicine* 176(8): 932-7.

Boeing Current Market Outlook 2012-2031. Boeing Commercial Airplanes 2012.

Congressional Budget Office. Long-Term Implications of the 2013 Future Years Defense Program. July 2012. Available at: www.cbo.gov/sites/default/files/cbofiles/attachments/07-11-12-FYDP_forPosting_0.pdf

Cyber Vision 2025: United States Air Force Cyberspace S&T Vision 2012-2025. United States Air Force Chief Scientist (AF/ST) Report. AF/ST-TR-12-01, 13 December 2012.

DARPA 2012; Patt, Daniel R.; Presentation: *Time to Market: On Drivers and Importance of Defense Systems Fielding Time.*

Department of Defense Strategy for Operating in Cyberspace. July 2011.

Department of Defense Directive 3000.09, Autonomy in Weapon Systems. http://www.dtic.mil/whs/directives/corres/pdf/300009p.pdf

Defense Science Board Study on Technology and Innovation Enablers for Superiority in 2030. 2013. http://www.acq.osd.mil/dsb/tors/TOR-2012-03-15-Summer_Study_2012.pdf

Effective Warfighting in Contested Environments (EWICE). Air Combat Command. 2013. Rew, W. and Fender, R. et al.

Energy Horizons: United States Air Force Energy S&T Vision 2011-2026. United States Air Force Chief Scientist (AF/ST) Report. AF/ST-TR-11-01-PR, 31 December 2011.

GAO 12-375. *DoD Supply Chain: Suspect Counterfeit Electronic Parts Can Be Found on Internet Purchasing Platforms.* GAO Report 12-375, Feb 21, 2012.

Global Trends 2025: A Transformed World. The National Intelligence Council. 2008.

Joint Operating Environment (JOE). 2010. U.S. Joint Forces Command.

Joint Operational Access Concept (JOAC). V1.0, 17 January 2012. Department of Defense.

Joint Strategy Assessment 2008-2028. Defense Intelligence Agency.

Joint Publication (JP) 3-12 *Cyberspace Operations* Final Coordination. 10 April 2012.

Joint Publication (JP) 3-13, *Information Operations*, 27 Nov 2012, http://www.dtic.mil/doctrine/new_pubs/jointpub_operations.htm

Lifting Off: Implementing the Strategic Vision for UK Aerospace. Aerospace Growth Partnership. 2013. https://www.gov.uk/government/publications/lifting-off-implementing-the-strategic-vision-for-uk-aerospace

McConnell, J. Michael (Director of National Intelligence). "Unclassified Statement for the Record from Testimony on Intelligence Community Annual Threat Assessment before Senate Armed Services Committee." Washington, DC. 27 Feb 2008. http://www.dni.gov/testimonies/20080227_testimony.pdf

Mission Impact of Foreign Influence on DoD Software. Washington, DC. Sep 2007. www.acq.osd.mil/dsb/reports/2007-09-Mission_Impact_of_Foreign_Influence_on_DoD_Software.pdf

National Security Strategy, May 2010. President of the United States.

NASIC System Threat Assessment Report, Aug 2009, "Global Hawk".

NASIC System Threat Assessment Report, January 2011, "MQ-9A Reaper".

Northrop, Linda. 2006. Ultra-Large-Scale Systems: The Software Challenge of the Future. Software Engineering Institute, Carnegie Mellon, Pittsburgh, PA.

NSA Information Assurance Directorate, 13 July 2010, "Operational Security Doctrine for the KOV-35 and KOV-35A Communication, Navigation, Identification, Processors (CNIP)"

NSF. 2010. Preparing the Next Generation of STEM Innovators: Identifying and Developing Our Nation's Human Capital" May 5, 2010. National Science Foundation (NSB-10-33) www.nsf.gov/nsb/publications/2010/nsb1033.pdf

Perrin, Chad. February 2010. "The danger of complexity: More code, more bugs". www.techrepublic.com/blog/security/the-danger-of-complexity-more-code-more-bugs/3076

Quadrennial Defense Review (QDR). 2010

Report of the Defense Science Board Task Force on Department of Defense Policies and Procedures for the Acquisition of Information Technology, March 2009. DTIC Report ADA498375.

Sustaining U.S. Global Leadership: Priorities for 21st Century Defense. January 2012.

Technology Horizons: A Vision for Air Force Science & Technology 2010-2030. Volume 1. United States Air Force Chief Scientist (AF/ST) Report. AF/ST-TR-10-01-PR, 15 May 2010.

"Trustworthy Cyberspace: Strategic Plan for the Federal Cybersecurity Research and Development Program". The White House Office of Science Technology and Policy. December 2011.

United Kingdom Ministry of Defense Global Strategic Trends out to 2040.

United States Air Force Strategic Environmental Assessment (SEA) 2010-2030. March 11, 2011. Directorate of Strategic Planning, Headquarters, United States Air Force (AF/A8X) 1070 Air Force Pentagon, Washington, DC 20330-1070.

"Worldwide Threat Assessment of the United States Intelligence Community for the House Permanent Select Committee on Intelligence - Unclassified Statement for the Record" Director of Nationl Intelligence. 2 February 2012.

"World Economic Outlook", International Monetary Fund. April 2011.

17. Acronyms

A2/AD	Anti-Access, Area-Denial
ADS-B/C	Automatic Dependent Surveillance-Broadcast/Contract
ADVENT	ADaptive Versatile ENgine Technology
AEHF	Advanced Extremely High Frequency
AF	Air Force
AF SAB	Air Force Scientific Advisory Board
AFMC	Air Force Materiel Command
AFLCMC	Air Force Life Cycle Management Center
AFRL	Air Force Research Laboratory
AFSPC	Air Force Space Command
AMC	Air Mobility Command
AOC	Air Operations Center
APT	Advanced Persistent Threat
ASD (R&E)	Assistant Secretary of Defense for Research and Engineering
ATC	Air Traffic Control
AWACS	Airborne Warning and Control System
BDA	Battle Damage Assessment
BLOS	Beyond Line of Sight
CAOC	Combined Air Operations Center
CCMD	Combatant Command
CMOS	Complementary Metal Oxide Semiconductor
COTS	Commercial Off-The-Shelf
C&A	Certification and Accreditation
CNE	Computer Network Exploitation
CAF	Combat Air Forces
C2	Command and Control
C2 and ISR	Command, Control, Intelligence, Surveillance and Reconnaissance
CONOPS	concept of operations
DARPA	Defense Advanced Research Projects Agency
DCIS	Data Confidentiality & Integrity Systems
DCGS	Distributed Common Ground System
DEW	Directed Energy Weapons
DIB	Defense Industrial Base
DINO	DoD Information Networks Operation
DoD	Department of Defense
DOE	Department of Energy
DON	Department of Navy
DSB	Defense Science Board
DT&E	Developmental Test and Evaluation
EW	Electronic Warfare
FAA	Federal Aviation Administration
FFRDC	Federally Funded Research and Development Center
FPGA	Field-Programmable Gate Array
FLOP	FLoating-point OPeration
FME	Foreign Military Exploitation
GIG	Global Information Grid
GPS	Global Positioning System

HAF	Headquarters Air Force
HAIPE	High Assurance Internet Protocol Encryptor
HEETE	Highly Efficient Embedded Turbine Engine
HRL	Hughes Research Laboratory
IBM	International Business Machines
IC	Intelligence Community
ICS	Industrial Control Systems
IOC	Initial Operational Capability
IPT	Integrated Product Team
IR&D	Independent Research and Development
ISR	Intelligence, Surveillance, and Reconnaissance
IT	Information Technology
ITV	In-Transit Visibility
ITAR	International Traffic in Arms Regulations
JCTD	Joint Capability Technology Demonstration
JOAC	Joint Operational Access Concept
JSF	Joint Strike Fighter
JSTARS	Joint Surveillance and Target Attack Radar System
JTAC	Joint Terminal Attack Controller
KPP	Key Performance Parameter
LIDAR	Light Detection And Ranging
LEO	Low Earth Orbiting
LRE	Launch and Recovery Element
MAJCOM	Major Command
MAF	Mobility Air Forces
MEF	Mission Essential Function
MIMO	Multiple-In Multiple-Out
MIT	Massachusetts Institute of Technology
NASA	National Aeronautics and Space Administration
NATO	North Atlantic Treaty Organization
NSA	National Security Agency
NSS	National Security Strategy
OCO	Offensive Cyberspace Operations
OSTP	White House Office of Science and Technology Policy
OT&E	Operational Test and Evaluation
PCPAD	Planning and Direction, Collection, Processing and Exploitation, Analysis and Production, and Dissemination
PNT	Position, Navigation and Timing
POTUS	President of the United States
RDT&E	Research, Development, Test and Evaluation
R&D	Research & Development
RFI	Request for Information
RFP	Request for Proposal
RPA	Remotely Piloted Aircraft
SA	Situational Awareness
SAF	Secretary of the Air Force
SIGINT	Signals Intelligence
SOF	Special Operations Forces
S&T	Science and Technology

S&TI	Scientific and Technical Intelligence
SMC	The Space and Missile Systems Center
SSA	Space Situational Awareness
STEM	Science, Technology, Engineering and Mathematics
SWAP	Size, Weight and Power
T&E	Test and Evaluation
TRL	Technology Readiness Level
TTPs	Tactics, Techniques, and Procedures
TTCP	The Technical Cooperation Program
UK	United Kingdom
US	United States
AF	United States Air Force
USCYBERCOM	United States Cyber Command
USD(AT&L)	Under Secretary of Defense for Acquisition, Technology, and Logistics
WMD	Weapons of Mass Destruction

18. *Global Horizons* Team

The following individuals contributed to the shaping the Air Force *Global Horizons* S&T vision and strategy through insights, counsel and functional expertise.

- ■ **Executive Leadership**
  - Honorable Michael Donley (SAF/OS), Secretary of the U.S. Air Force
  - General Mark Welsh, (AF/CC), Chief of Staff of the U.S. Air Force
  - Honorable Jamie Morin (SAF/US), Acting Undersecretary of the U.S. Air Force
  - General Larry Spencer (AF/VC), Vice Chief of Staff
  - Gen William Shelton (AFSPC/CC), Cyberspace Superiority Core Function Lead Integrator
  - Gen Mike Hostage (ACC/CC), Air Superiority Core Function Lead Integrator
  - Gen Janet Wolfenbarger (AFMC/CC), Agile Combat Support Core Function Lead Integrator
  - Gen Herbert Carlisle (PACAF/CC)
  - Gen Philip Breedlove (AFE/CC)
  - Lt Gen James Kowalski (AFGSC/CC)

- ■ **Senior Governance Team**
  - Dr. Mark Maybury (Chair) (AF/ST), Chief Scientist of the U.S. Air Force
  - Lt Gen John Hyten (AFSPC/CV)
  - Lt Gen Larry James (AF/A2)
  - Lt Gen Mike Basla (SAF/CIO A6)
  - Lt Gen Michael Moeller (AF/A8)
  - Dr. Jacqueline Henningsen (AF/A9)
  - Maj. Gen. Garrett Harencak (AF/A10)
  - Lt Gen Thomas Travis (AF/SG)
  - Lt Gen Charles Davis (SAF/AQ)
  - Lt Gen Thomas Owen (AFLSC/CC)
  - Maj Gen Neil McCasland (AFRL/CC)

- ■ **Key Stakeholders**
  - **MAJCOMs**
  - Lt Gen William Rew (ACC/CV)
  - Lt Gen Douglas Owens (AETC/CV)
  - Maj Gen Everett Thomas (AFGSC/CV)
  - Lt Gen Robert Allardice (AMC/CV)
  - Lt Gen Andrew Busch (AFMC/CV)
  - Lt Gen John Hyten (AFSPC/CV)
  - Maj Gen George Williams (AFSOC/CV)
  - Lt Gen Stanley Kresge (PACAF/CV)
  - Maj Gen Noel Jones (AFE/CV)
  - Maj Gen Craig Gourley (AFRC/CV)
  - Brig Gen James Witham (ANG/CV)

    Air Staff
  - Lt Gen Frank Gorenc (AF/CVA)
  - Lt Gen Darrell Jones (AF/A1)
  - Lt Gen Burton Field (AF/A3/5)
  - Lt Gen Judith Fedder (AF/A4/7)
  - Dr. David Walker (AF/AQR)
  - Lt Gen James Jackson (AF/RE)

    Secretariat
  - Mr. Charles Blanchard (SAF/GC)
  - Ms. Marilyn Thomas (SAF/FM)

- Mr. Terry Yonkers (SAF/IE)
- Ms. Kathleen Ferguson (SAF/IE)
- Ms. Heidi Grant (SAF/IA)
- Mr. David Tillotson (SAF/US(M))
  Domain Experts
- Lt Gen Ellen Pawlikowski (SMC/CC)
- Maj Gen Samuel Greaves (AFSPC/A8/9)
- Maj Gen Earl Matthews (AF/A3C/A6C)
- Maj Gen Ken Merchant (AAC)
- Maj Gen Robert Otto (AFISRA/CC)

■ *Global Horizons* **Area Study** <u>Leads</u>**, Co-Leads and Key Members**
  Core Function Teams
- Threat: <u>Gary O'Connell (NASIC)</u>, Maj Gen Jim Keffer (AF/A2), Col Matthew Hurley (AF/A2DD)
- Air: <u>Dr. Don Erbschloe (AMC/ST)</u>, Dr. Dave Robie (ACC/ST), Bill Harrison (AFRL/RQ), Dr. Bob Peterkin (AFRL/RD), Dr. Mikel Miller (AFRL/RW), Dr. Kamal Jabbour (AFRL/RI), Barth Shenk (AFRL/RQ), Lt Tom Mock (AFRL/RQ)
- Space: <u>Dr. Doug Beason (AFSPC), Dr. Jim Riker (AFRL/RV)</u>, Col Scott Beidleman (SMC/XR), Dr. Roberta Ewart (SMC/XR), Dr Alan Weston (NASA)
- Cyber: George <u>Duchak/Dr. Rich Linderman (AFRL/RI),</u> Dr. Doug Beason (AFSPC), Arthur Wachdorf (24AF), Frank Konieczny (SAF/A6 CTO), Mike Kretzer (688th), Steve Schneider (AFRL/RY), Dr. Rusty Baldwin (AFIT/ENGE)
- C2 and ISR: <u>Dr. Steven K. Rogers (AFRL/RY/RI), Dr. Terry Wilson (AFRL/RY)</u>, Mr. Stan Newberry (AFC2IC), Dr. Chris Yeaw (AFGSC/ST), Jeff Eggers (AF/A2), Keith Hoffman (NASIC), Mr. Bill Marion (ACC)
- Mission Support (Acquisition, T&E, Workforce): <u>Dr. David Walker (SAF/AQR),</u> Susan Thornton (AFMC/EN), Col Derek Abeyta (AF/TE), Maj Mike Dunlavy (SAF/AQR), Lt Col Dan Ward (LCMC), Ed Kraft (AEDC/CZ), Dr. Alok Das (RY)
- Enabling Technology: <u>Dr. Jennifer Ricklin (AFRL), Dr. Chuck Matson (AFRL/AFOSR/CL)</u>, Dr. Pat Carrick (AFRL/AFOSR/RT)
  Global Sector Teams
- Manufacturing and Materials – <u>Dr. Barry Farmer (AFRL/RX), Doug Bowers (AFRL/RQ)</u>, Dr. Mikel Miller (AFRL/RW), Col Keith Bearden (AFLCMC/XZ), Rollie Dutton (AFRL/RXM)
- Transportation and Logistics - <u>Dr. Don Erbschloe (AMC/ST)</u>, Steven Hofmann (JPDO, Next Gen), Lt Col Jerry Hollman (AFSFC/SFOZ), Lt Col Scott Spiers (AFSFC/SFOZ), Bob Nagel (AMC/A8XC), Sonja Glumich (AFRL/RIGA)
- Energy, Utilities & Mining – <u>Dr. Kevin Geiss (SAF/IEN)</u>, Dr. Bill Harrison (AFRL/RZ), Dr. Bob Peterkin (AFRL/RD); Lt Col Charles Bulger (SAF/IEN)
- Health Care & Pharmaceutical – <u>Dr. Morley Stone (AFRL/RH)</u>, Dr. Deb Niemeyer (59 MDW/ST), Lt Gen Tom Travis (AF/SG); Col Randy Ashmore (AFMSA/SG5)
- Communications, Information Technology, Financial Services - <u>George Duchak/Dr. Rich Linderman (AFRL/RI),</u> Dr. Doug Beason (AFSPC), Dr. Kamal Jabbour (AFRL/RI), Dr. Paul Antonik (AFRL/RI), Dr. Rob Gold (ASD R&E), Dr. Emily Krzysiak (AFRL/RI)
- Education and Training – <u>Dr. Bruce Murphy (AU/VP Academic Affairs)</u>, Dr. Todd Stewart (AFIT), Dr. Nathaniel Davis (AFIT), Dr. Morley Stone (AFRL/RH), Dr. John Geis (AU/AFRI), Dr. Steven Hansen (AU), BGen Scott Vander Hamm/Craig Seeber (AETC/A5/8/9A), Lt Col Chris Bohn (AETC/Spaatz Center), Dr. Aaron Byerley (AFA)
  Study Management
- Study Management and Leadership: Col James Greer (AF/ST)
- Study Administration Support: Penny Ellis (AF/ST)

■ **Additional Subject Matter Experts:**
- Mr. Randall Walden (SAF/AQI), Dr. Mark Gallagher (A9), Linda Millis (DNI, Private Sector Partnerships), Col Rex R. Kiziah (AFSPC/ST)

19. Senior Independent Expert Reviewer Group

■ **Senior Independent Expert Review Group**
  - **Air:**
    - Natalie Crawford[6], Senior Fellow, RAND
    - Dr. Tom Hussey, former AFOSR
    - Prof Mark Lewis[3], IDA
    - Lt Gen George Muellner[6], (Ret) AF
    - Robert Osborne, NNSA
    - Dr. Jaiwon Shin, NASA
  - **Space**
    - Keith Hall[2], Booze Allen Hamilton
    - Don Kerr[2]
    - Matt Linton, NASA ARC-IS
    - Dr. David Miller[6], MIT, Vice Chair AF SAB
    - Dr. Mason Peck, NASA CTO
    - Dr. Rami Razouk[6], Senior Vice President, Aerospace
    - Brig Gen (ret) Pete Worden, NASA
    - Dr. Michael Yarymovych[3, 6]
  - **Cyber**
    - Alan Bernard, MIT LL
    - Dr. Steve Bussolari, MIT LL
    - Prof Ed Feigenbaum[3], Stanford
    - Tim Grance, NIST
    - Lt Col Marion Grant, USCYBERCOM/J9
    - Glenn Gafney, CIA
    - Gen (ret) Mike Hayden[1], AF
    - Paul Laugesen, NSA/TAO
    - Dr. Boyd Livingston, DoD
    - Andrew Makridis, CIA
    - Lt Gen (ret) Ken Minihan[4], AF
    - Dr. Paul Nielsen, Director and CEO, Software Engineering Institute
    - Dr. Larry Schuette, ONR
    - Dr. Mike Wertheimer, DoD
    - Dr. Yul Williams, NSA/CSS TOC
    - Dr Starnes Walker, FltCyber, Navy
    - Dr. Steven King, OSD(R&E), Chair, Cyberspace Priority Steering Committee
    - Mr. Gil Vega, DOE
  - **C2 and ISR**
    - Lt Gen (ret) Ted Bowlds, USAF
    - Dr. Steve Cross, GTRI
    - Lt Gen (ret) David Deptula, USAF
    - Lt Gen (ret) Robert Elder, USAF
    - Mr. Al Grasso, MITRE, President and CEO
    - Mr. Ray Haller, MITRE
    - Dr. Jim Hendler, RPI
    - Maj Gen (ret) Ken Israel, USAF
    - Prof Alex Levis[3], GMU
    - VADM (ret) Mike McConnell1, USN
    - Dr. Donna Rhodes, MIT SEAri
    - Dr. Ralph Semmel, JHU-APL, Director

  - **Mission Support**
    - Mr. Norm Augustine, former chair Lockheed Martin

- Mr. Giorgio Bertoli, Army
- Landon Derentz (DoE)
- Dr. Tom Ehrhard, OSD(P)
- Mr. John Gilligan[5]
- Mr. Brian Hughes, AT&L
- Gen (Ret) Duncan McNabb[7], USAF
- Dr. Tim Persons, GAO Chief Scientist
- Mr. Alan R. Shaffer, OSD (ASD R&E)
- Ms. Heidi Shyu, ASA(ALT)
- Dr. Harold Gregory Smith, NGA
- Mr. Ben Steinberg, DoE
- Dr. Steve Walker, Deputy Director, DARPA

- **Enabling Science and Technology**
  - Charles Bouldin, NSF
  - Gen (Ret) Mike Carns[7]
  - Stan Chincheck, NRL
  - Prof. Werner Dahm[3], Director Security & Defense Systems Initiative (SDSI), Arizona State Univ
  - Dr. Peter Friedland, formerly NASA, AFOSR Advisor
  - Dr. David Honey, DNI
  - Mr. Terry Jaggers, NAS
  - Leland Jameson, NSF
  - Dr. Walter Jones, ONR
  - Dr. Paul Kaminski, DSB Chair
  - Richard Matlock, MDA
  - Gen (Ret) Jim McCarthy, USAF
  - Dr. Kathryn Sullivan, NOAA
  - Tomas Vagoun, NITRD
  - Konrad Vesey, IARPA
  - Lauren Van Wazer, OSTP
  - Prof. Patrick H. Winston, MIT
- **Coalition**
  - Air Vice Marshall Brecht, RAF, UK
  - Dr. Brian Hanlon. DSTO, Australia
  - Mr. Simon Kippin, RAF, UK
  - Mr. Christopher McMillan, MoD, Canada
  - Mr. Philip Rayburn, British Embassy, Washington D.C.
  - Dr. Anthony Shellhase, Australian Embassy, Washington D.C.
  - Norbert Weber, MoD, Germany

Notes:
[1] Former Director of National Intelligence
[2] Former Director of the National Reconnaissance Office
[3] Former Chief Scientist of the AF
[4] Former Director of NSA and DIA
[5] Former AF Chief Information Officer
[6] AF SAB Executive Committee
[7] Former AF VCSAF

20. *Global Horizons* **Terms of Reference**

Background

Global demographic, economic, technological, and military trends forecast an increasingly complex, competitive, and contested future. An Air Force wide S&T vision is needed to articulate a path forward that anticipates future threats, mitigates vulnerabilities, and shapes and takes maximal advantage of impending and unexpected opportunities. In collaboration with joint, interagency, and international partners, this study will create an integrated, Air Force-wide, near-, mid-, and far-term S&T vision that identifies revolutionary capabilities to sustain our strategic advantage and assure Global Vigilance, Global Reach and Global Power across air, space, and cyberspace.

Approach

Partnering with the air staff, MAJCOMs, internal stakeholders, and external organizations, AF/ST will:

- Identify and forecast global trends (e.g., economic, demographic, S&T, military) and S&T revolutions that may radically transform threat vectors and/or opportunity spaces

- Identify global opportunities (including weak signals) that promise to dramatically change cost structures in acquisition (e.g., agile manufacturing), human talent (e.g., automation), operations (e.g., process change), sustainment, and/or revolutionize human/system performance (e.g., new materials, nanotechnology, robotics).

- Identify and prioritize the most promising S&T areas in air, space, and cyberspace where the AF, with its strategic S&T partners, should lead, follow, and/or watch in the near (present-FY17), mid (FY18-22) and far (FY23-27) terms.

- Prioritize the most strategic AF problems and identify best practices (e.g., partnerships, competitions, prizes) for motivating solutions that help overcome obstacles and achieve more rapid and economical S&T advancement.

- Coordinate regularly with AF leadership and via periodic updates to SAF/OS and AF/CC.

Products

- Preliminary *Global Horizons* S&T vision to AF leadership by 1 June, 2013.

- Final briefing to SAF/OS, AF/CC, SAF/US and AF/CV by 15 August 2013.

- Final report by 1 October 2013 articulating global trends, S&T game changers, and most promising near-, mid- and far-term vectors.

www.ingramcontent.com/pod-product-compliance
Lightning Source LLC
Chambersburg PA
CBHW081308180526

45170CB00007B/2614